European Federation of Corrosion
Publications

NUMBER 5

A Working Party Report

Illustrated Case Histories
of Marine Corrosion

Published for the European Federation of Corrosion

by The Institute of Metals

THE INSTITUTE OF METALS

1990

Book Number 496

Published in 1990 by The Institute of Metals
1 Carlton House Terrace, London SW1Y 5DB

and
The Institute of Metals
North American Publications Center
Old Post Road, Brookfield VT 05036

British Library Cataloguing in Publication Data

Illustrated case histories of marine corrosion.
1. Corrosion by seawater
I. IJsseling, F. P. II. Institute of Metals, 1985- III.
Series
620.11223

ISBN 0-901462-86-1

Library of Congress Cataloging in Publication Data

Available on application

Data processing and text design by *P i* c A Publishing Services,
Drayton, Nr Abingdon, Oxon

Colour origination by Chroma Graphics, Singapore

Made and printed in Great Britain by Wincanton Litho, Wincanton, Somerset

Contents

Chapter 3 *Specific Cases*...**26**

Heat Exchangers   *84*

European Federation of Corrosion
Publications
Series Introduction

The EFC, incorporated in Belgium, was founded in 1955 with the purpose of promoting European co-operation in the fields of research into corrosion and corrosion prevention.

Membership is based upon participation by corrosion societies and committees in technical Working Parties. Member societies appoint delegates to Working Parties, whose membership is also expanded by co-option of other individuals.

The activities of the Working Parties cover corrosion topics associated with inhibition, education, reinforcement in concrete, microbial effects, hot gases and combustion products, environment sensitive fracture, marine environments, surface science, physico-chemical methods of measurement, the nuclear industry, and computer based information systems. Working Parties on other topics are established as required.

The Working Parties function in various ways, e.g. by preparing reports, organising symposia, conducting intensive courses, and producing instructional material, including films. The activities of the Working Parties are co-ordinated, through a Science and Technology Advisory Committee, by the Scientific Secretary.

The administration of the EFC is handled by three Secretariats: DECHEMA in Germany, the Société de Chimie Industrielle in France, and the Institute of Metals in the United Kingdom. These three Secretariats meet at the Board of Administrators of the EFC. There is an annual General Assembly at which delegates from all member societies meet to determine and approve EFC policy. News of EFC activities, forthcoming conferences, courses etc. is published in a range of accredited corrosion and certain other journals throughout Europe. More detailed descriptions of activities are given in an occasional Newsletter prepared by the Scientific Secretary.

The output of the EFC takes various forms. Papers on particular topics, for example, reviews or results of experimental work, may be published in scientific and technical journals in one or more countries in Europe. Conference proceedings are often published by the organisation responsible for the conference.

In 1987, the Institute of Metals was appointed as the official EFC publisher. Although the arrangement is non-exclusive and other routes for publication are still available, it is expected that the Working Parties of the EFC will use the Institute of Metals for publication of reports, proceedings etc. wherever possible.

A. D. Mercer
EFC Scientific Secretary
Institute of Metals London, UK

EFC Secretariats are located at:

Mr R. Wood
European Federation of Corrosion
The Institute of Metals
1 Carlton House Terrace
LONDON SWl Y 5DB
UK

Dr D. Behrens
Europäische Föderation Korrosion
DECHEMA
Theodor-Heuss-Allee 25
D-6000
FRANKFURT (M)
Germany

M R. Mas
Fédération Europeene de la Corrosion
Société de Chimie Industrielle
28 Rue Saint-Dominique
F-75007 PARIS
FRANCE

Acknowledgements

This report has been prepared by the European Federation of Corrosion Working Party on Marine Corrosion, the membership of which is as follows:

H. Arup	*Denmark*	F. P. IJsseling	*Netherlands*
C. Christensen	"	O. Steensland	*Norway*
P. Drodten	*Germany*	A. Domanski	*Poland*
E. Hargarter	"	B. Wallèn	*Sweden*
W. Semrau (guest)	"	J. Tavast	"
O. R. W. Forsèn	*Finland*	L. Svensson	"
S. Yläsaari	"	K. Härtel	*Switzerland*
L. Lemoine	*France*	R. O. Muller	"
A. M. Beccaria	*Italy*	J. Parker	*UK*
G. Rocchini	"	B. Todd	"

Foreword

It is a great pleasure to present this book, containing a collection of specific cases of marine corrosion damage in seawater handling systems. This collection has been assembled by a sub-committee of the working party on Marine Corrosion of the European Federation of Corrosion, in which the following members have collaborated actively:

C. Christensen

R.O. Muller

J.G. Parker

E.B. Shone

J. Tavast

B. Todd

F.P. IJsseling

We thank these members for spending their time and making available their knowledge. Particular mention should be made of the contribution by E.B. Shone, the former chairman of the working party. In addition to initiating this project, he acted as the focal point for the work and collated many of the cases cited. We also thank the other members of the working party for their support and criticism and other persons who have contributed by providing documented cases of corrosion damage.

It is very important for the engineers in the field to be able to recognize corrosion damage when it occurs. Approximately 50% of the failures encountered in practice are due to some mechanical aspect, the other 50 % being caused by corrosion. In about 50% of all failures the cause can be attributed to some initial error, e.g. inappropriate choice of materials, incorporation of undesirable design features, etc.

The purpose of this book is to illustrate the majority of types of corrosion damage, which may arise in seawater handling systems. It is hoped that the examples presented will enable the engineer in the field to be able to recogise the cause of component failure and to apply appropriate remedial measures.

The cases of corrosion damage have been made available by different sources. The sub-committee being dependent on the material offered, not all cases which are found in practice could be covered, nor should the number of cases dedicated to a specific failure be taken as a measure for the severity of the problem.

F.P. IJsseling
Chairman of EFC Working party
on Marine Corrosion

Chapter 1

Introduction

1.1 Purpose

The aim of this book is to provide engineers with an elementary review of cases of corrosion damage which may occur during marine service, and advise on possible remedial action. The book has been limited to seawater handling systems. In the introductory chapter background information is given on the theory of marine corrosion, the nature of seawaters, the different corrosion types which can be encountered in practice and the possible remedial measures. In the next chapter elementary information is provided on the several classes of materials which are used as construction materials in marine systems. This is followed by the main part, a review of cases of corrosion damage which have occurred in practice in seawater handling systems.

1.2 Theory of Marine Corrosion

All aqueous corrosion processes including those taking place in marine environment are basically of an electrochemical nature, in which a metal reacts with its environment, in this case seawater. The corroding metal passes into the seawater as positively charged ions (oxidation or anodic reaction) which can be expressed by the following chemical reaction formula:

$$M \rightarrow M^{n+} + ne \qquad (1)$$

The ionisation of the metal releases electrons which participate in the balancing reaction in which electrons are consumed (reduction or cathodic reaction). Thus, the anodic and cathodic processes, both involving the exchange of electrons, occur at the boundary between the metal surface and the corrosive environment. Both processes can be visualised by the passage of electric currents through the boundary plane: from the metal to the solution in the anodic case and in the opposite direction for cathodic processes (Fig. 1). The cathodic processes most relevant in the context of corrosion are due to dissolved oxygen and hydrogen ions:

$$O_2 + 2 H_2O + 4e \rightarrow 4 OH^- \qquad (2)$$

$$\text{and} \quad 2H^+ + 2e \rightarrow H_2 \qquad (3)$$

The corrosion reaction of iron, for example, can be expressed chemically as follows:

$$2 Fe \rightarrow 2 Fe^{2+} + 4e \qquad (4)$$

$$O_2 + 2 H_2O + 4e \rightarrow 4 OH^- \qquad (5)$$

$$2 Fe + O_2 + 2 H_2O \rightarrow 2 Fe^{2+} + 4 OH^- \qquad (6)$$

This is a typical example of an electrochemical corrosion process in which the rate of anodic metal dissolution is balanced by the rate of cathodic reduction of dissolved oxygen at the metal surface (Fig. 1). The rates of both electrochemical oxidation (anodic) and reduction (cathodic) reactions depend in principle on temperature, concentration and potential. Generally higher temperatures and concentrations of the reacting species promote higher reaction rates. The effect of the potential is on the rate of electron exchange, increasing potentials leading to an increase of the oxidation rate and a decrease of the reduction rate. A decrease of the potential has the opposite effect. Theoretically it can be derived, as is found in many practical cases, that the logarithm of the current is linearly dependent on potential. In the corroding state the combination of metal oxidation and the concurrent reduction reaction proceed at the same potential, the free corrosion potential,

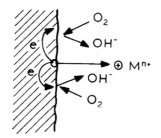

Figure 1: Schematic representation of electrochemical corrosion process involving oxidation of metal M to metal ions M^{n+} and coupled with reduction of dissolved oxygen.

which can be measured by means of a reference electrode (e.g. silver / silver chloride, saturated calomel, etc.) and an appropriate voltmeter with high input impedance ($>10^8\ \Omega$).

Generally the following corrosion cases can be distinguished:

(1) the corrosion rate is governed primarily by the rate of electron exchange at the metal surface and consequently by all factors which influence the value thereof, the potential being particularly important , as explained before;

(2) the corrosion rate is governed either by the supply of the constituent to be reduced to or by the transport of metal ions away from the surface; this case - which often applies to marine corrosion - is characterised by a strong dependence of the corrosion rate on the rate of transport in the solution (diffusion and convection);

(3) the corrosion rate can also be influenced by the electrical conductivity of the solution or of an insulating layer on the metal surface: the greater the resistance the lower the corrosion rate;

(4) a special case is passivity, where the metal surface is covered with a thin layer which acts as a barrier to the transport of ions and/or electrons, thereby lowering the metal oxidation rate to a very low value.

The electrochemical processes and the dependency of their rates on potential can be visualized by means of polarisation curves, which are an algebraic summation of the separate anodic and cathodic currents. A further discussion of this aspect, however, is outside the scope of this introduction.

An important observation is that corrosion unlike mechanical strength is not a specific alloy property. In addition to the nature of the environment a number of other variables may influence the corrosion process in a quantitative as well as in a qualitative way (e.g. temperature, flow velocity, etc.). The dependency on so many variables which frequently interact with the corrosion process makes corrosion less easy to predict.

Another important distinction is between uniform and localized corrosion processes. The former applies to metal corrosion where the corrosion attack proceeds uniformly over the metal surface in contact with the environment, due to the conditions in the metal and in the solution being sufficiently homogeneous (Fig. 2a). This is the most usual type of attack, the rates of which are generally fairly well known and therefore can be relatively easily incorporated in the design.

On the other hand, in localized corrosion processes the oxidation of the metal proceeds at specific sites due to the presence of heterogenities (Fig. 2b). In a number of cases of localised corrosion the location of the attack is mainly determined by the metallurgical structure of the metal. Examples are intergranular corrosion and selective corrosion. In other cases constructional aspects of the structure may play an important part as for example in bimetallic corrosion, crevice corrosion, flow and/or stress induced corrosion.

Local corrosion processes can proceed on a microscopic as well as on a macroscopic scale. Generally the number of variables governing the rate of such localized processes is greater than with uniform corrosion, leading to increased uncertainty whether a localized process will proceed, and if so, at what rate. For this

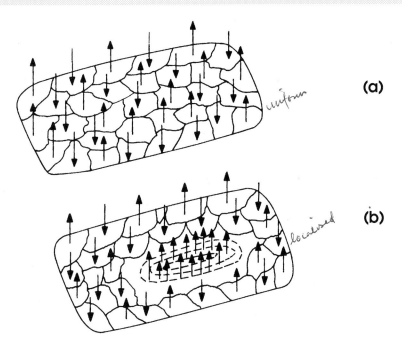

Figure 2: Difference between uniform (a) and localised (b) corrosion. In the first case the anodic and cathodic currents are distributed homogeneously over the metal surface. In the second case the oxidation reaction occurs in a fixed position, while the coupled cathodic reactions proceed in a surrounding area.At a large distance from the anodic site the oxidation and reduction currents again balance.

reason localized corrosion processes are more unpredictable and dangerous and as a small local perforation or a crack in a component can lead to outage of the whole system and thus their technological impact is larger.

In some cases the distinction between uniform and local attack is not very sharp. It is possible to distinguish several forms of localised corrosion, ranging from areas of shallow attack to the formation of very narrow and deep pits or cracks.

Factors which often play an important part in localized attack are:

- the cathode/anode area relationship
- differential aeration cells
- pH at the cathodic and anodic sites
- corrosion products or other layers present on the metal surface or formed during the corrosion reaction
- active-passive cells.

1. 2. 1 Cathode/anode area
It can be derived that the local attack will be more pronounced when the cathodic area is larger than the anodic area and when the rate of the cathodic process is higher (Fig. 3). Under these conditions a larger cathodic current is available to support the anodic reaction. The balance between anodic oxidation and cathodic reduction is that between the electrical charge allied to these processes. When the anodic process is confined to a relatively small area a large anodic current density results, which is directly proportional to the corrosion rate and thus metal loss.

1. 2. 2 Differential aeration cells
Frequently a situation occurs where part of a metal surface is in contact with a solution in which oxygen is dissolved while another part of the same construction is in contact with the solution having a lower concentration of oxygen. Mostly such situations arise by limitation of the transport of the oxygen to different parts of the surface. It can be shown that at the surface which is in contact with the lower oxygen concentration,

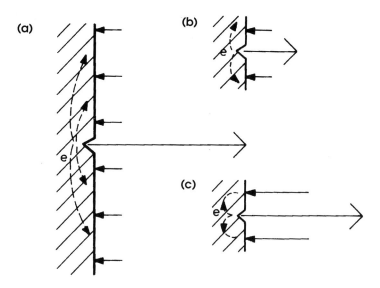

Figure 3: In electrochemical corrosion the electric charge allied to the anodic process is equal to the charge allied to the cathodic process. Thus a larger cathodic surface will give rise to a larger anodic current, i.e. a larger corrosion rate (compare a and b), as will larger cathodic current densities (compare b and c).

the corrosion rate when coupled to a surface in contact with a higher oxygen concentration increases as compared with the uncoupled state. Contrarily, for the surface in contact with the higher oxygen concentration the opposite effect is observed on coupling (Fig. 4).

1. 2. 3 pH changes during corrosion attack
In the majority of corrosion cases either dissolved oxygen or H^+ (acid) are involved in the corrosion reaction. In both cases the cathodic reaction involves an increase of the pH (i.e. decrease of acidity) at the metal surface, as can be deduced from eqs. 2 and 3. On the contrary the oxidation of the metal at the anodic site frequently leads to a decrease of the pH (acidification), for example by direct formation of H^+ during the electrode reaction:

$$\text{e.g. Fe} + H_2O \rightarrow \text{Fe(OH)}^+ + H^+ + 2e \qquad (7)$$

The actual change at the surface will depend on the solution chemistry (possible buffering action) as well as on the local flow conditions. However, it is clear that in cases where the anodic and cathodic sites are separated from each other (occluded cells: crevices, pits, cracks) locally substantial pH-changes may occur (Fig. 5).

1. 2. 4 Corrosion products and deposits
When corrosion products or surface layers are present the formation of local cracks and pits in the layers can give rise to local corrosion at the base of these defects. A well-known example is mill scale on steel, in which cracks can develop due to different thermal expansion of the scale and the underlying steel. These cracks result in the formation of local corrosion cells, consisting of the bases of the cracks, i.e. the underlying steel surface, as the anodic sites and the mill scale surface as the cathodic site. When a deposit is present on a metal surface an anodic zone is often initially formed by a crevice between the deposit and the metal surface. Depending on the progress of the reaction the anodic zone may spread underneath the deposit (Fig. 6).

1. 2. 5 Active-passive cells
When dealing with the corrosion aspects of passive alloys frequently situations occur when part of the alloy surface is in the desired passive state with low corrosion rate, while part of the surface has lost its passive state and has become activated (depassivated), and thus exhibits a much higher corrosion rate. A passive surface

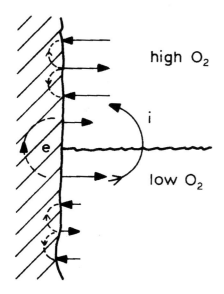

Figure 4: Schematic representation of a differential aeration cell showing the increase of anodic current at the lower oxygen zone in the vicinity of the boundary between the surface areas in contact with respectively low and high oxygen concentrations. At the high oxygen side the opposite effect occurs and the anodic current will decrease. Depending on geometrical conditions and electrical conductivity of the solution at larger distances from the boundary the anodic and cathodic reactions will balance again.

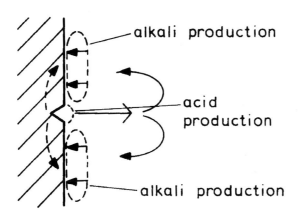

Figure 5: Schematic representation of tendency to decrease and increase of pH at respectively predominantly anodic and cathodic sites.

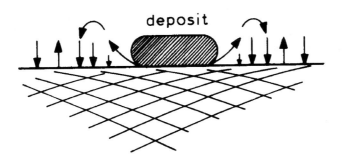

Figure 6: Schematic representation of predominantly anodic reaction under deposit on the metal surface.

has the inherently higher corrosion potential and both surfaces are internally short-circuited through the metal. The effect of such an active-passive cell is to enhance the corrosion rate at the active part of the alloy surface (Fig.7). The magnitude of the enhancement depends on the ratio between the active and passive surface areas, the characteristics of the electrochemical reactions and the potential difference.

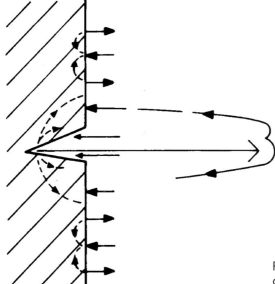

Figure 7: Schematic representation of active-passive cell.

However, in many cases the above mentioned factors interact, which may complicate the understanding of the mechanism of localized corrosion.

Following this, necessarily very limited, introduction to some of the general aspects of localised corrosion, those corrosion types which are mainly of interest within the framework of marine corrosion will be discussed in section 1.4.

1.3 Nature of Seawaters

Seawater can be characterised as a dynamic system, consisting of an aqueous solution of salts and gases, also containing organic and living material and insoluble particles. In relation to corrosion processes the following properties are relevant:

- the large salt content of about 35 grams/kg, of which 90 % is sodium chloride;

- the dissolved salts generate a low electric resistivity of the seawater, which is therefore a good electrolyte, thus the corrosion rate is primarily determined by the rate of the electrochemical processes at the metal surface (activation) and possible transport of active matter in the solution phase (diffusion/convection); also when local corrosion occurs its rate generally will be enhanced because a larger cathodic surface is available for supporting the anodic reaction at the corroding site;

- the dissolved carbon dioxide leads to the formation of carbonic acid, which undergoes ionization to bicarbonate and carbonate ions; the resulting system and in addition the presence of undissociated boric acid provides a constant and relatively high pH (7.8-8.3) to the seawater;

- due to the high pH and consequently low availability of hydrogen ions the principal species which is reduced at the metal surface is dissolved oxygen; however, in crevices and other restricted areas the oxygen concentration may become very low so other oxidizing species may become involved, e.g. hydrogen ions and bacteriological components;

- the high chloride concentration generally will enhance corrosion, e.g. by increasing the rate of the metal oxidation process and/or by provoking local breakdown of passive layers;

- organic compounds generally will be detrimental in enhancing the corrosion rate by their complexing properties;

- calcium and magnesium compounds may often be involved in the formation of calcareous layers which may be advantageous from the standpoint of decreasing the oxygen supply to the surface and thereby decreasing the corrosion rate; however, they may be detrimental by decreasing the rate of heat exchange by deposition on the metal surface;

- a large amount of living biological matter will tend to accumulate and grow on exposed surfaces; apart from the increase of weight and resistance to flow the decaying products may produce components like ammonia and sulphur compounds which may provoke corrosive action; moreover due to biochemical oxidation the oxygen concentration at the metal surface may become very low, giving rise to bacteriological processes, e.g. involving sulphate reducing bacteria, which can enhance the corrosion process, even in the absence of dissolved oxygen.

The composition of seawater from the open seas and oceans is remarkably constant regarding the major components. However, there are some deviations, as for instance the Baltic Gulf due to the large influx of river water and the Red Sea due to the large amount of evaporation. Generally the corrosion processes are governed and/or influenced by the minor compounds and the dissolved gases and these tend to fluctuate much more, in particular in coastal areas where most of the specific marine applications are to be found. The main reason for this is the general formation of layers, for example of a calcareous nature, corrosion products, slimes and other bioactive material. The corrosion processes depend to a large extent on the properties of such films and it is in this context that the main differences between corrosion attack in different areas is to be found. The dissolved oxygen concentration, for example, depends on temperature and salinity, but also on chemical processes like photosynthesis (production of oxygen) and biochemical oxidation (consumption of oxygen). As a result the oxygen concentration may fluctuate with depth, location and season. At coastal sites the salinity may be lower due to dilution with inland water and generally the content of organic and biological material will be larger due to the increased amount of nutrients and the large surface available for settlement. An important additional factor is the generally large pollution at coastal areas due to industrial, domestic and agricultural waste products. As a result of these effects the corrosivity of seawater at coastal sites may differ significantly as compared with that obtained from test results in clean, undiluted seawater from the open seas; thus "on the spot" testing is necessary. Test results obtained with stored and/or recirculated seawater or synthetic mixtures must also be regarded with suspicion due to differences between their corrosivity and that of real seawater.

1.4 Types of Corrosion

Due to the many variables which may affect the corrosion process several different types of corrosion may occur. Classification by appearance (uniform or localized) is most helpful for a preliminary discussion and will be used here. However, in this respect it must be stated that nomenclature to classify the different corrosion types is not very logical. A number of corrosion types are named according to the phenomenon which cause the corrosion, e.g. erosion-corrosion and crevice corrosion. On the other hand a number of corrosion types are named according to their appearance after corrosion, e.g. pitting and intergranular corrosion. This inconsistency leads to the fact that in some corrosion types, e.g. erosion-corrosion, pits can be formed, although the corrosion effect is significantly different from classical pitting corrosion. In the following paragraphs those corrosion types, which are paramount in marine corrosion systems, will be discussed.

1. 4. 1 Uniform attack
Uniform attack over large areas of a metal surface is the most common type of corrosion. According to the electrochemical mixed potential theory the anodic and cathodic processes are both distributed homogeneously

over the metal surface (Fig. 8). For this type of corrosion to occur it is required that the corrosion system exhibits no major heterogeneities, either in the metal microstructure and the environment (concentration differences) or in the exposure conditions (e.g. temperature differences, irregular flow patterns, local stresses). Large differences can be found between the uniform corrosion rates of different corrosion systems.

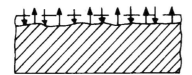

Figure 8: Uniform attack.

1.4.2 Bimetallic corrosion (also named galvanic - and contact corrosion)

This type of corrosion occurs when two (or more) dissimilar metals, which are electrically connected to each other, are in contact with the same corrosive environment. In such cases the metal with the most negative corrosion potential in the uncoupled state (the active member of the couple) will show enhanced corrosion as compared with the corrosion rate in the uncoupled state but the other metal (the more noble metal with the higher corrosion potential in the uncoupled state) will corrode less on coupling (Fig. 9). Dissimilar metals in the uncoupled state generally will differ in corrosion potentials and corrosion rates. In the coupled condition only one mixed potential can be attained which is situated between the original potentials of the separate metals.

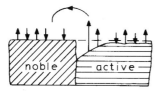

Figure 9: Bimetallic corrosion.

For the active metal this means an increase of the potential, which is accompanied by an increase of the corrosion current, whilst the potential of the noble metal will be lowered and as a consequence so will the corrosion rate.

As in the case of differential aeration and active-passive cells a local current can be defined as the surplus current leaving the active metal and entering the more noble component at the metal-solution interface. Generally the following factors will influence the bimetallic corrosion process:

- the difference between the corrosion potentials of the uncoupled metals;
- the electrochemical polarization characteristics of the metals in the vicinity of their corrosion potentials;
- the electrical resistance of the bimetallic circuit, which can be attributed to the conductivity of the solution and/or corrosion products or paint films on the surface; the effect of the resistance is to lower the galvanic current;
- the surface areas of the cathodic and anodic zones;
- geometrical considerations regarding the distribution of current across the boundary between the metals, for example the distance between the anode and the cathode.

Generally, a large difference in corrosion potentials, small slopes for the partial electrochemical reactions, a large ratio of the cathodic/anodic surface areas and a small distance between the metals will be detrimental. As a general guide of the possible effect of bimetallic corrosion tables are available, listing the corrosion potentials of separate metals and alloys in a given environment under specified conditions (galvanic series). As free corrosion potentials may depend on the chemical composition of the environment, galvanic series are pertinent to a given corrosive environment. As in example in Fig. 10 the galvanic series for seawater is shown. However, it must be noted that in this particular case the free corrosion potentials of some classes of alloys,

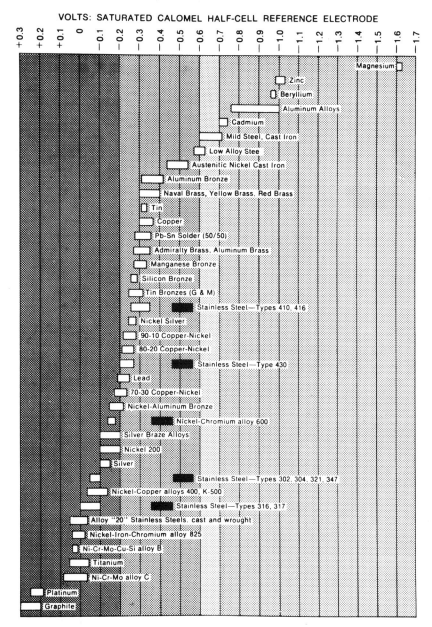

VOLTS: SATURATED CALOMEL HALF-CELL REFERENCE ELECTRODE

Alloys are listed in the order of the potential they exhibit in flowing sea water. Certain alloys indicated by the symbol: ▇ in low-velocity or poorly aerated water, and at shielded areas, may become active and exhibit a potential near −0.5 volts.

Figure 10: Galvanic series in flowing seawater (2.5 - 4 m/s), temperature range 11 - 30°C.

Source: Marine Corrosion, Causes and Prevention. John Wiley & Sons Inc., New York.

i.e. stainless steels, nickel-alloys and titanium, may become substantially higher in practical applications due to the formation of bacteriological slime layers. Also the free corrosion potentials may depend on factors such as flow velocity and temperature.

1. 4. 3 Flow-induced corrosion

The effect of electrolyte flow is generally to enhance corrosion due to the improved mass transfer of ionic or solid anodic reaction products away from, or cathodic reaction species or intermediates to, the metal surface. Under laminar and uniform steady turbulent flow conditions the resulting accelerated corrosion is generally of a uniform nature, provided the diffusion layer is also uniform.

In many systems the corrosion rate will be limited by the mass transfer that may be involved in the anodic or cathodic partial reactions. However, with the higher flow rates associated with turbulence raisers, which are inevitable under real conditions, locally higher levels of turbulence are achieved giving rise to a localized corrosion termed erosion-corrosion. Apart from increasing the mass transfer process by the increased turbulence removal of protective layers of corrosion product may also occur, further accelerating the corrosion process. Moreover, if the corrosion potential of a corrosion product scale covered metal surface is more noble than that of the bare surface, the potential of the latter is increased, accelerating the anodic process still further (Fig.11). If the electrolyte also contains entrained gas bubbles, which impinge against the metal surface at a specified location, the same kind of accelerating effect may be obtained, which is termed impingement attack. Impingement attack can also be associated with abrasion due to the impact of solid particles and is then termed erosion-corrosion or sometimes abrasion corrosion. Cavitation corrosion, which arises due to the collapse of vapour cavities or bubbles at a solid surface may also be classified within this group. Cavitation may give rise to material loss ranging from predominantly mechanical to predominantly electrochemical mechanisms. At low intensities of cavitation with vapour bubbles collapsing in the electrolyte the effect on corrosion may be the same as that due to impingement attack. However, at higher intensities of cavitation, bubbles collapsing on the metal surface may totally remove the corrosion product layer, while at even higher cavitation levels the forces exerted are sufficient to mechanically deform and erode the surface. Erosion corrosion is characterised in appearance by grooves, gullies, waves, rounded holes and valleys and usually exhibits a directional pattern. In particular with impingement attack a very specific pattern is observed, known as horse-shoe attack (Fig.12). The nature and properties of the protective layers that form on most metals and alloys are very important from the standpoint of resistance to the afore mentioned corrosion forms. The ability of these layers to protect the metal depends on the speed or ease with which they form when originally exposed to the environment, their resistance to mechanical damage or wear and their rate of reforming when destroyed or damaged. A hard, dense adherent and continuous layer would provide better protection than one that is easily removed by mechanical means or worn off. A brittle scale that cracks or spalls under stress may not be protective. Frequently the nature of the protective layer that forms, depends upon the specific environment to which it is exposed and this determines its resistance to erosion-corrosion by that fluid.

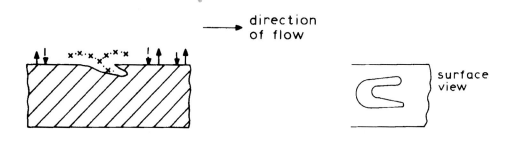

Figure 11: Erosion-corrosion.   Figure 12: Horse-shoe attack.

1. 4. 4 Crevice corrosion

Crevice corrosion is a type of intense localized corrosion frequently occuring within crevices and other shielded areas on metal surfaces exposed to corrosive liquids (Fig.13). This type of attack is usually associated with small volumes of stagnant solution caused by holes, gasket surfaces, lap joints, and crevices under bolt and rivet heads.

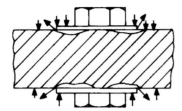

Figure 13: Example of crevice corrosion.

Crevice corrosion is characterised by a geometrical configuration in which the cathodic reactant (usually dissolved oxygen) can readily gain access by convection and diffusion to the metal surface outside the crevice, whereas access to the layer of the stagnant solution within the crevice is far more difficult and can be achieved only by diffusion through the narrow mouth of the crevice. For crevice corrosion to occur the crevice must be wide enough to permit entry of the solution but sufficiently narrow to maintain a stagnant zone of solution within the crevice, limiting the transport processes of diffusion and migration of ions. Solutions containing chloride ions are the most conducive to crevice corrosion. The mechanism of crevice corrosion is fairly complicated. The first step is the formation of a differential aeration cell due to cathodic reduction of dissolved oxygen within the crevice, access of fresh oxygen to the crevice solution being very limited. In contrast, the oxygen concentration at the freely exposed zone outside the crevice will be subtantially higher. Due to the action of the differential aeration cell the anodic dissolution of the metal within the crevice will continue, resulting in the accumulation of positive metal ions within the crevice solution. The excess positive charge will be balanced, partly by positive ions migrating and diffusing to the outside, partly by diffusion of negative ions (Cl^-) to the inside. The net effect is an increase of the metal chloride concentration within the crevice, which on hydrolysis of the metal ions produces hydrogen ions:

$$Fe^{2+} + 2\,H_2O \;\rightarrow\; Fe(OH)_2 + 2\,H^+ \tag{9}$$

The $Fe(OH)_2$ is not protective, while the liberation of H^+ results in a decrease of the pH to values varying between 3 and 0.5, depending on the metal ions involved and the time elapsed. According to this mechanism an incubation period can be distinguished, necessary for the formation of a corrosive solution inside the crevice, characterised by almost zero oxygen content and high metal chloride and acid concentrations.

When this stage has been reached the anodic process within the crevice will proceed rapidly, which may be partly attributed to the direct reduction of H^+ inside the crevice. So the incubation period is followed by a propagation period, in which the rate of the anodic process frequently is seen to increase with time (autocatalytic process). In the case of stainless steels, which are particularly prone to crevice corrosion, the rate of the anodic dissolution is further accelerated by the relatively large driving potential which is formed between the outside of the crevice, which is passivated readily, and the inside which is depassivated by the aggressive crevice solution.

Corrosion processes which are closely related to crevice corrosion are deposit attack, in which a crevice is formed between a metal surface and a deposit (sand, mud, dirt), and filiform corrosion. The latter corrosion form is characterised by the formation of a network of threadlike filaments of corrosion products on the surface of a metal coated with a transparent lacquer or a paint film, as a result of exposure to humid atmosphere.

1. 4. 5 Pitting corrosion

Pitting is a form of localised attack that may result in perforation of the metal (Fig. 14). These holes may be small or large in diameter, but in most cases they are relatively small. Pits are sometimes isolated or so close together that they look like a rough surface. Generally a pit may be described as a cavity or hole with the surface diameter about the same as, or less than, the depth. Pitting is one of the most destructive and insidious forms of corrosion. It causes equipment to fail because of perforation with only a small percent weight loss of the entire structure. Pitting can often be ascribed to well-defined heterogenities associated with the metal-environment system (e.g. discontinuities in mill scale, precipitates or protective layers). However, pitting can also occur in systems, which are apparently free from heterogeneities, provided aggressive ions such as Cl⁻ are present. Pitting corrosion occurs frequently at pores or damaged parts in:

- non-conductive layers on the metal surface;

-metallic surface layers which are noble to the base metal: bimetallic corrosion will occur, leading to pits in the base metal. In this latter case the pitting process can develop fairly rapidly, leading to deep pits within a short time.

Moreover, undercutting of the metallic surface layer may occur under organic layers, e.g. paint; or corrosion product films and scales. Pitting is often promoted by the potential of the film covered surface being somewhat more noble than the bare surface.

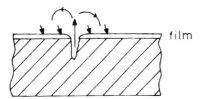

Figure 14: Example of pitting corrosion.

Pitting corrosion is often found with stainless alloys. Crevice corrosion and pitting of stainless steels have a number of features in common and it has been stated that pitting may be regarded as crevice corrosion in which the pit forms its own crevice. However, whereas a macroscopic heterogeneity determines the site of attack in crevice corrosion, the sites of attack in pitting are determined by microscopic or sub-microscopic features in the passive film.

Factors such as temperature and flow conditions will also influence the pitting mechanism to a great extent. At higher temperature the susceptibility of passivated alloys to pitting generally increases. Stagnant conditions will also be detrimental owing to the fact that the adverse solution conditions which develop in the micropit are not swept away thereby hampering a possible repassivation.

1. 4. 6 Selective corrosion

In this corrosion form, which is also called selective leaching or parting, one element, generally the most active one, is selectively removed from a solid alloy (Fig.15). As a result the components of the alloy react in proportions which differ from their proportions in the alloy. Apart from the general term the process is often named after the removed element in specific cases, e.g. dezincification of brasses, dealuminification of certain Al-bronzes, etc.

A very well-know example of selective corrosion is dezincification of brass. Dezincification is readily recognized as the alloy assumes a red copper colour, i. e. in contrast to the original yellow. There are two types of dezincification: one is the uniform, or layer-type, and the other is the localised, or plug-type.

The dezincified zones are mechanically weak and porous. Stagnant conditions (formation of deposits and scales), higher temperatures, low oxygen content and the presence of sulphides in the seawater generally will promote dezincification. The metal structure and composition are also important: two-phase brasses are more susceptible than the single phase types.

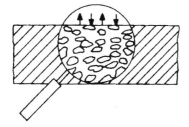

Figure 15: Selective corrosion.

The susceptibility to dezincification of brasses can be decreased by decreasing the Zn-content (for example, red brass containing 15% Zn is almost immune). The addition of 1% Sn to 70-30 brass is advantageous (Admiralty brass). Further improvement is obtained by adding small amounts of As, Sb or P to the alloy as inhibitor. However, such additions are not effecctive in high Zn β-alloys.

Another alloy, which is well-known for selective leaching, is grey cast iron. In this case the interconnected graphite flakes are noble to iron, forming an excellent galvanic element on a microscopic scale. The iron is dissolved, leaving a porous mass, consisting of graphite, voids and rust. As a consequence the cast iron loses its metallic properties as well as its strength (graphitisation). As in the case of brass, dimensional changes do not occur, and if not detected in time dangerous situations may develop.

1. 4. 7 Intergranular corrosion
A metal is composed of crystals or grains. Generally the grain boundaries of the metal will show some enhanced reactivity, due to the local disordered structure. In many applications this is of little or no consequence, because the increase of reactivity relative to the matrix is only limited. However, under certain conditions grain interfaces will become very reactive and intergranular corrosion results, involving localised attack at and adjacent to the boundaries, with relatively little corrosion of the grains themselves. The alloy disintegrates (grains fall out of the matrix) and/or loses its strength (Fig.16). The main causes of intergranular corrosion are:

(i) impurities at the grain boundaries,

(ii) enrichment of one of the alloying elements, or

(iii) depletion of one of these elements in the grain-boundary area.

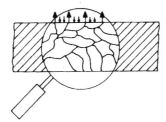

Figure 16: Intergranular corrosion.

A well-known example is the sensitisation of austenitic stainless steel type AISI 304 in the temperature range from 500 to 800°C. In this temperature range carbide precipitates, consisting of $Cr_{23}C_6$, are formed locally at the grain boundaries. The resulting effect is local depletion of chromium dissolved in the matrix. As the presence of dissolved chromium promotes the passivity of the alloy, the Cr-depleted grain-boundary zones will be susceptible to attack, because the dissolved chromium content locally is not sufficient to keep these zones within the passive range. On activation of the grain boundary zones active-passive cells will be formed, consisting of the grain surfaces in the passive and the grain boundaries in the active state. Due to the relatively large driving power of these cells and the unfavourable ratio of anodic/cathodic areas, rapid attack will be possible. A closely related phenomenon is weld-decay of austenitic stainless steels, which may occur after welding when the material adjacent to the weld is heated within the temperature zone in which sensitisation occurs .

1. 4. 8 Stress-corrosion cracking (SCC)

Cracks may be formed in many construction materials when they are exposed to a corrosive environment while a mechanical tensile stress is also present at a level which in itself should be not harmful. The cracks which are formed by this combined action are often difficult to detect and when allowed to grow may lead to sudden catastrophic failure (Fig. 17). In the SCC process two stages can be often distinguished: the first formation of the crack (initiation period) and the propagation of the crack (propagation period) ending with mechanical failure, resulting entirely from mechanical action on the reduced cross-sectional area. A necessary condition for SCC to occur is the presence of stress, which may arise from various sources, e.g. applied, residual, thermal. The environmental conditions for SCC to occur are rather specific, in the sense that not all environments promote cracking. Well-known materials which may show susceptibility to SCC in chloride environments are austenitic stainless steels containing Cr and Ni and a number of Al-alloys. Some copper alloys may be susceptible in the presence of ammonia. Cracking proceeds generally perpendicular to the applied stress. Both intergranular cracking, proceeding along grain boundaries, and transgranular cracking, advancing without apparent preference for grain-boundaries are observed. Intergranular and transgranular cracking also may occur in the same alloy, depending on the environment and/or the metal structure. Cracks can also vary in degree of branching. Generally the susceptibility to SCC increases with increasing temperature. For a number of alloy/environment combinations a safe temperature can be indicated, below which the susceptibility to SCC is practically nil. The progress of the SCC process under specified conditions of environment and stress level are also influenced by the potential. The interdependence of the variables in SCC, namely microstructure, electrochemistry and response to stress, support the suggestion that these interact in a variety of ways, leading to a continuous spectrum of mechanisms rather than a single mechanism.

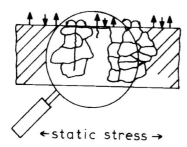

Figure 17: Stress-corrosion cracking.

←static stress →

1. 4. 9 Hydrogen embrittlement

High-tensile steels (low-alloy steels in which the strength is produced by suitable heat treatment) and, to a lesser extent, some high-strength alloys of copper, aluminium and titanium, are subject to SCC as a result of absorbed hydrogen. The phenomenon is so widespread as to constitute a special case of SCC, called hydrogen embrittlement. In addition to this there are many other cases of SCC, in which hydrogen is also thought to be involved in the reaction mechanism. It is apparent that for hydrogen embrittlement to occur hydrogen must be adsorbed at the metal-solution interface and that part of the adsorbed hydrogen must transfer across the interface to become absorbed by the metal. Hydrogen evolution and adsorption is favoured by low pH. It has been established that solution conditions inside a crack can be far more acid than in the bulk solution, which is ascribed to accumulation of metal salts (mainly chloride) and subsequent hydrolysis reactions in the crack. Possible sources of hydrogen in steel arise from such treatments as refining, welding, pickling, electroplating, phosphating and paint stripping. However, corrosion processes in which the corrosion reaction involves evolution of hydrogen, also arise. As stated above, even in the case of weakly alkaline solutions such as seawater, the solution inside a crack may become strongly acidic with consequent discharge of hydrogen. In all these cases the uptake of hydrogen by the metal is promoted when the recombination of H to H_2 is poisoned (for instance by sulphur compounds). The application of coatings, which are more active than steel (e.g. Zn, Al and Cd) may also be the cause of hydrogen embrittlement. The same applies to cathodic protection, either with sacrificial anodes or with an impressed current system.

1. 4. 10 Corrosion fatigue

This type of corrosion is attributed to the conjoint action of cyclic stress and a corrosive environment, resulting in cracks propagating from the surface in a direction perpendicular to the stress (Fig. 18). Fatigue is often characterized by means of the relation between the amplitude of the cyclic stress (S) and the number of cycles before failure (N). In this way often a S-N curve is obtained, showing a critical value of the stress fatigue limit, below which no fatigue will occur (Fig.19). The influence of the corrosive environment is to lower the fatigue limit, sometimes even to zero. In addition the crack growth rate may increase.

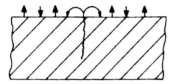

Figure 18: Corrosion fatigue.

←dynamic stress →

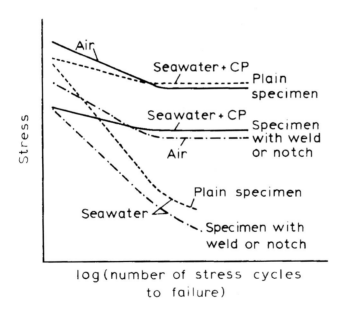

Figure 19: Schematic effects of seawater and cathodic protection on S-N curve of steel (also including effects of notches and welds).

Apart from the corrosivity of the environment, the frequency of the cyclic stresses are also important, low frequencies generally being more dangerous. As in the case of SCC and some other localized corrosion forms an initiation and a propagation period can often be distinguished, both periods being adversely affected by the corrosion reaction. As can be seen from Fig. 19 cathodic protection may be effective in overcoming the detrimental effects of cyclic stresses although when applied to high strength alloys the possibility of hydrogen embrittlement should be taken into account.

Set of civil leva lw

1.5 Corrosion Protective Measures

The general methods which can be applied to prevent or reduce corrosion attack, can be distinguished as follows:

(1) materials selection

(2) design

(3) pretreatment

(4) control or modification of environment

(5) application of coatings

(6) electrochemical (potential) control

The first three mentioned groups of preventive measures are fundamental and generally applicable to all corrosion systems. In materials selection due attention should be paid to possible corrosion reactions - which requires a thorough knowledge of the environmental conditions , possible mechanical stresses, and the mode of operation of the system. In particular the possibilities for the occurrence of local corrosion forms such as crevice attack, pitting, etc., have to be considered, as local perforations can lead to outage of a whole system. The choice of possible additional preventive measures has to be made in combination together with the materials selection procedure. Often the choice is between cheaper and more readily available materials which require more maintenance and/or earlier replacement or more expensive alloys which are intrinsically more corrosion resistant. In this context cost calculations should be made on a total life cycle cost basis, including initial cost as well as the cost of maintenance, additional preventive measures, repair, indirect cost due to outage, etc. In short all cost factors which are relevant to keep the system operational during its planned lifetime have to be considered.

Corrosion prevention by design is very important, as by implementing a number of relatively cheap and easy to apply measures the cost of maintenance and thereby the total cost can be reduced substantially. In the design stage subjects such as accessibility for inspection, maintenance and repair, type of joints, shaping of components, bimetallic corrosion, possible entrapment of water or moisture, etc. should be taken into account (Fig. 20). Situations to be avoided in piping include sharp bends and constrictions, turbulence raisers, crevices, accumulation of debris, dirt, mud, etc. The most general rule is to avoid heterogeneities in a system such as local stresses, temperature differences, different alloys, etc.

Of course, the use of different alloys cannot always be avoided, however it is possible by abiding to a few rules to limit the possible detrimental effects. Measures to be considered in this context are: a favourable ratio between the noble and the less noble parts, coating (either the whole surface around the contact or only the surface of the more noble component, never only the surface of the active component), matching the electrochemical properties of the alloys (free corrosion potential as well as polarisation behaviour) and electrical insulation. Parts of the system which are critical and liable to corrosion should be designed so that they can be easily removed. Of course, design considerations are closely related to materials selection. It is very important that corrosion prevention is considered in design and materials selection from the beginning and not as an after-thought. Corrosion prevention should be an integral part of the design.

Several possibilities exist for pretreatment of materials, for example:

(1) surface pretreatment for providing a clean, homogeneous surface as an improved basis for the growth of protective layers of corrosion products and the appreciation of externally applied coatings,

(2) passivating surface pretreatment to obtain a strong, homogeneous passive scale on the surface prior to the exposure to the seawater, and

(3) a heat treatment of the alloy, either to obtain a microstructure which is advantageous from the standpoint of possible corrosion attack or to remove possible mechanical stresses.

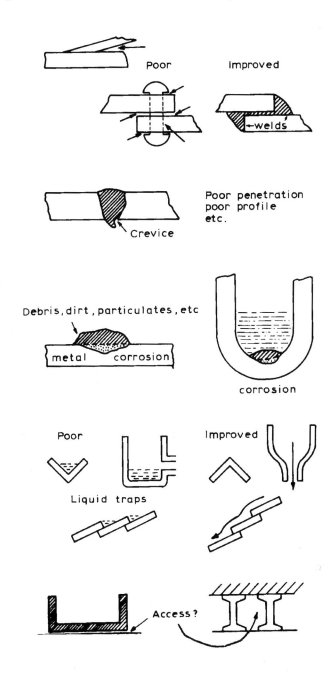

Figure 20: Examples of the possible influence of design on the susceptibility to corrosion of constructions.

The possibilities for control and modification of the environment are closely related to the operation of the system. Measures to be considered are regulation of pH and removal of dissolved oxygen. The addition of inhibitors to the environment can also be classified in this category, although a number of inhibitors can also be regarded as substances leading to coating of the alloy surface. Obviously in closed systems much more can be done in this respect than in once through systems. Examples in marine applications are to be found in the seawater desalination industry and in the addition of certain components to seawater cooling systems (for example $FeSO_4$, copper- and aluminum-ions and organic compounds with the last type being mostly used prior to practical use of the system).

The protection against corrosion by the application of coatings is widely accepted in the marine field (Fig. 21). Coatings can be classified into a number of groups:

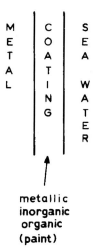

metallic
inorganic
organic
(paint)

Figure 21: Corrosion protection by the application of coatings.

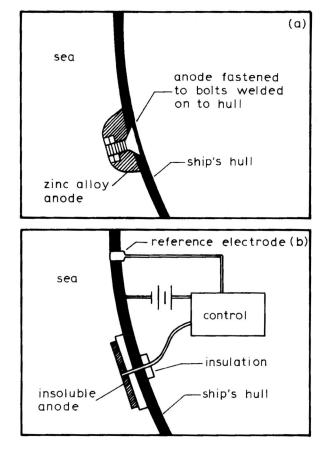

Figure 22: Cathodic protection of a ship's hull by means of: (a) sacrificial anodes, (b) impressed current.

(1) metallic (e.g. aluminium, zinc or aluminium-zinc alloy on steel);

(2) inorganic (e.g. vitreous enamel and conversion coatings as aluminium oxide on aluminium, phosphate on steel and chromate on steel and non-ferrous alloy types);

(3) organic (e.g. bituminous systems, synthetic resins);

(4) paint systems, either acting as a barrier between the metal and the seawater to exclude water, oxygen and ionic substances, or complemented in many cases by the incorporation of pigments which prevent or retard corrosion. Active metal powder, e.g. aluminium or zinc, can similarly be added. If necessary for the specific application components can also be added to increase the mechanical strength (e.g. reinforcement by glass fibres) and/or to increase the path length for detrimental compounds to reach the metal surface (e.g. glass flakes). For all kinds of application of protective layers an adequate surface preparation is required.

Another widely applied method of corrosion protection is cathodic protection which consists of control of the oxidation rate of the metal by control of the potential. In cathodic protection the potential of the system to be protected is decreased from the free corrosion potential in a negative (cathodic) direction, so that the corrosion rate of the metal is decreased. The method can be accomplished either by galvanic coupling of the system to be protected with anodes of a less noble alloy (sacrificial anodes: Fig.22a) or by supplying a direct current in the appropriate direction (impressed current method: Fig. 22b). The sacrificial method is rather straightforward with low capital investment, however, sufficient anode material is required for long life otherwise periodical replacement is necessary. The impressed current method is somewhat more complicated, requiring a supply of direct current (transformer/rectifier), generally combined with a control system to regulate the current output to control the potential at the required level, permanent anodes to pass the current, a reference electrode and cabling. However, notwithstanding the higher initial cost there are quite a number of applications where impressed current cathodic protection is more cost-effective within a few years of operation. Cathodic-protection is preferably combined with a protective coating to limit the current demand, in which case some extra demands are required to be met by the coating, e.g. to withstand the negative potential without blistering. As sacrificial anodes zinc and aluminium can be used for steel and steel can be used for copper alloys.

There are also applications of cathodic protection to stainless alloys, the main aim, however, being to prevent the onset of pitting and crevice corrosion. Examples of cathodic protection are found in shipping, offshore constructions, pipelines on the bottom of the sea, condensers, etc.

1.6 References

General textbooks on corrosion and the prevention of corrosion, frequently including chapters on specific marine corrosion problems:

1. Corrosion, vol. 1 and 2; ed. L. L. Shreir. Published by Newnes-Butterworths, London/Boston, 2nd edn., 1976.

2. Corrosion Engineering; M.G. Fontana and N. D. Greene. Published by McGraw-Hill, New York, 1967.

3. Corrosion Basics. Published by National Association of Corrosion Engineers, Houston, Tx., U. S. A., 1984.

4. Introduction to Corrosion Prevention and Control; P. J. Gellings. Published by Delft University Press, 1985.

5. Metals Handbook, 9th edn., vol .13. Published by ASM International, Metals Park, OH, U. S. A., 1987.

6. Design and Corrosion Control; V. R. Pludek, Published by MacMillan Press, Ltd., London, 1977.

7. Handbuch des kathodischen Korrosionsschutzes; W. v. Baeckmann & W. Schwenk, Verlag Chemie GmbH; 2nd. edn., 1989.

8. Cathodic Protection: J. Morgan. Published by National Association of Corrosion Engineers, 2nd edn. Houston, Tx, U. S. A., 1987.

9. Ship Painting Manual; A. M. Berendsen, Graham and Trotman, London, 2nd ed., 1989.

10. ISO 8044, 1986-12-15: Corrosion of metals and alloys-Terms and definitions.

Textbooks and papers dedicated to marine corrosion aspects:

11. Marine Corrosion; F. L. LaQue. Published by JohnWiley & Sons, Inc., New York, 1975.

12. Marine and offshore corrosion; K. A. Chandler. Published by Butterworths, London, 1985.

13. J. C. Rowlands and B. Angell: Corrosion for marine and offshore engineers; Marine Engineering Practice, vol.2, part 11, The Institute of Marine Engineers / Marine Media Management Ltd., London, 1976.

14. J. F. Deegan: Corrosion in ships; Special Ships, April 1979, $\underline{4}$.

15. L. Kenworthy and J. C. Rowlands: Recent developments in preventing marine corrosion; M.E.R. May 1979, $\underline{7}$.

General textbooks and papers on sea water:

16. Introduction to Marine Chemistry: J. P. Riley and R. Chester. Published by Academic Press, London, 1971.

17. S. C. Dexter: in Section on Marine Corrosion, Metals Handbook, 9th edn., vol. 13, 893, 1987. Published by ASM International, Metals Park, OH, U. S. A., 1987.

18. S.C. Dexter and C. Culberson: Global Variability of Natural Seawater, Mater. Perform., 1989, $\underline{20}$, (9), 16.

Other pictorial collections of corrosion damage:

19. Corrosion Atlas: E.D.D. During. Published by Elsevier, Amsterdam, 1988.

20. Das Gesicht der Korrosion: A.F. Bertling and E.A. Ulrich, Publ. Techn. uberwachungs - Verein Bayern e.v., München, 1969.

Chapter 2

Materials Used in the Marine Environment

2.1 Carbon Steel and Low Alloyed Steel

Carbon steel is the most commonly used construction material due to its advantageous mechanical properties, ease of handling and low initial cost. Unprotected carbon steel will suffer uniform corrosion in seawater at a rate of 0.1-0.6 mm/year under steady-state conditions. However, as the water velocity increases the corrosion rate will increase and erosion-corrosion occurs at water speeds in excess of 1 m/s. Additional corrosion forms are: bimetallic corrosion when coupled to more noble metals, pitting under deposits and in mud zones, and corrosion fatigue. Usually this class of materials is protected by coatings, cathodic protection or by the combination of both. When used without protection a large corrosion allowance is required, which may be unacceptable due to increased weight. By addition of small amounts of alloying elements such as nickel, copper and aluminium low alloyed steels have been developed with a higher strength and somewhat increased corrosion performance. However, these HSLA steels still require additional protection, albeit less on a life cycle cost basis. Another development is the use of high strength steels (HY 80-130 type) for special purposes as for example submarine hulls. Apart from the corrosion forms already mentioned corrosion cracking and hydrogen embrittlement should also be taken into account for high strength steel.

2.2 Cast Irons

Cast iron has been used as a standard material of construction for pump and valve bodies, water boxes, pipe fittings and other cast components. However, due to its tendency to graphitisation, a form of selective corrosion in which the iron dissolves and a very brittle graphitic structure remains, the material has been replaced in many cases by more corrosion resistant and/or higher strength materials. Examples of these are ductile cast iron, nickel cast iron containing 1-3% Ni, and in particular, austenitic nickel cast irons, containing 14-32% nickel, 1.75-5.5 % chromium and up to 7% copper (Ni-Resist).

2.3 High Alloyed Steel

The high-alloyed steels, often termed corrosion resistant or stainless steels, owe their corrosion resistance to the formation of a passive surface film. This is a severe restriction as these steels only show a good service performance when the passive film remains undamaged over the whole metal surface, or when any damage is quickly healed.

As soon as local perforation of the passive film or depassivation occurs, active-passive cells are formed which promote heavy local corrosion. For this reason a warning is appropriate against the inconsiderate application of stainless steels in general and in particular in seawater. There are many grades and kinds of stainless steels, which are not suitable at all for marine service without additional protective measures. So conventional austenitic stainless steels such as AISI 304 and 316 have a strong tendency to several forms of localized corrosion such as pitting and crevice corrosion in chloride containing waters. This is a severe limitation of the application of these steel types in seawater service, unless a number of conditions is met, including continuous contact with clean flowing seawater, regular surface cleaning and seawater temperature not above ambient. If these conditions are met in practice frequently good service experience is obtained. However, in a number of cases satisfactory service is due to the inadvertent or deliberate application of cathodic protection by coupling to carbon steel. However, due to modern developments in steel fabrication, by which the microstructure and the chemical composition (in particular with regard to the content of C, S, P and N) can be much better controlled, a number of steel types have evolved which are very useful for seawater applications. These alloys show high strength, ductility, workability and weldability, and moreover their resistance to localized corrosion has been increased significantly. The elements chromium, molybdenum and nitrogen if present improve the resistance to pitting, which may be indicated by the widely used formula:

$$PRE = \%Cr + 3.3 \times Mo\% + 16 \times N\%$$

The factor PRE is called the pitting resistance equivalent and should be above at least 40 for reliable seawater applications, depending on seawater as well as service conditions. Three types of alloys have evolved:

(1) the fully austenitic high-alloyed stainless steels of which the types containing 6% or more Mo are the most resistant to pitting and crevice corrosion. These steels have shown excellent seawater service in condensers and piping systems;

(2) the ferritic high-alloyed types are used when a very good resistance against stress corrosion cracking is required together with a good pitting/crevice corrosion resistance;

(3) the duplex stainless steels have a microstructure consisting of an about 1:1 mixture of austenite and ferrite and combine the near-immunity to SCC in seawater of the ferritic stainless steels with the toughness and the ease of practical application of the austenitic types. Duplex steels are available both as wrought and cast products, resulting in a number of applications including gate valves, pump shafts, cast propellers, seawater piping, seawater-inert gas scrubber systems and critical seal components of nuclear submarines.

2.4 Nickel-base Alloys

As with the stainless steels the corrosion performance of nickel alloys is strongly dependant upon the formation of a passive surface film , capable of self-repair. The Monel*-type alloys, containing about 30% copper sometimes with small additions of metals such as Fe, Mn, and Al, are widely used. The Monels* exhibit a general corrosion performance in seawater that is somewhat superior to most of the copper base alloys but are generally somewhat less resistant to crevice and pitting corrosion in quiet seawater. Other high-performance nickel alloys contain up to about 20% Cr, 20-30% Fe and up to 15% Mo or 15-20% Cr and up to 15% Fe. Within these ranges the Hastelloy* alloy types C, C-276 and C22 are found, and also Inconel* 625 and Incoloy *825. These alloys combine very good corrosion properties, i.e. low general corrosion, good resistance to pitting, crevice corrosion and SCC, with good mechanical properties. For this reason these alloy types are frequently used for critical components, notwithstanding their relatively high price. Typical seawater applications include propeller blades, bellows, expansion joints and shaft seals.

2.5 Aluminium Alloys

Aluminium and aluminium alloys are generally resistant to corrosion due to the formation of a protective film of Al-oxide. However, generally it is advisable to produce such a protective film artificially by anodizing in combination with a sealing process. Moreover the application of a protective coating can be envisaged and/ or cathodic protection. In the latter case potential control is required to avoid excessive cathodic values which otherwise could lead to local alkalisation which is to be avoided because of the amphoteric character of aluminium. For marine applications the following alloy types are used:

(1) non-heat treatable wrought alloys, containing 2-5.5% Mg possibly with some Mn;

(2) heat-treatable alloys containing 0.5-1.5% Mn, 0.4-1.3% Si and possibly some Mn and Cr, and

* Trade marks

(3) the cast alloys with 3-11% Mg or 3-13% Si or combinations thereof, while to the Mg-containing types Mn is often added.

Regarding localized corrosion — aluminium is a very active element and it will often be used in practice coupled with more noble alloys and therefore the danger of bimetallic corrosion has to be kept in mind.

Some alloy types are liable to pitting and/or intergranular corrosion, the latter due to potential differences between precipitates and the surrounding microstructure. Regarding pitting corrosion the presence of copper and/or copper ions is detrimental as copper ions precipitate on the aluminium surface, giving rise to extended pitting. Some higher strength alloys are also sensitive to SCC. Typical applications in the marine field are deckhouses on ships, helicopter landing decks on offshore platforms and hulls of smaller ships. Other uses are in the desalination field.

2.6 Copper Alloys

Copper and in particular copper alloys are used extensively in seawater due to their naturally occurring and protective corrosion-products and good anti-fouling characteristics. Due to the formation of protective layers the corrosion rate decreases significantly on first exposure to fresh seawater, the long-term steady state corrosion rate in seawater being in the range 0.01-0.025 mm/yr. A point to be considered in materials selection is their sensitivity towards flow-induced corrosion. A great number of copper alloys have been developed, some with only slight compositional variations. The main alloy groups can be given as follows:

- brasses	(Cu + Zn)
- bronzes	(Cu + Sn)
- aluminium bronzes	(Cu + Al)
- gunmetals	(Cu + Sn + Zn)
- cupronickels	(Cu + Ni)

However, this is a fairly general first order division, as usually small quantities of other elements are added to obtain alloys with specific desired properties for improved corrosion resistance and/or mechanical strength. For marine applications the single-phase alpha brasses CuZn29Sn1 (Admiralty brass) and CuZn22Al2 (aluminium brass) are in use and in both cases 0.02-0.05% arsenic is added to prevent dezincification. Examples of the two-phased alpha-beta brasses are Muntz metal (CuZn40) and naval brass (CuZn40Sn1). Dezincification may be a problem in these alloys, which are mainly used for thick-walled components such as tube plates. Another type is manganese bronze, which is basically a CuZn40 alloy with additions of elements such as tin, iron, manganese, nickel or aluminium to increase the mechanical performance.

Of the bronzes the phosphor bronze (CuSn8P0.4) is widely used in marine engineering and the shipbuilding industry. AP-bronze (CuSn8Al1Si0.1) has been developed specifically for use in polluted seawater. The aluminium bronzes proper contain 5-11% Al, possibly with additions of up to 6% Fe, 3% Mn, 7% Ni or 3% Si. Control of composition and microstructure is very important in this class of alloys, which are available in both wrought and cast forms. Some types are sensitive to selective corrosion, more specifically dealuminification. As in the case of stainless steels for marine applications the type and grade of the Al-bronze should be selected carefully, as many compositions are not suitable for seawater service. The gunmetals are basically alloys of Cu, Sn and Zn, often with the addition of lead to improve casting characteristics. Examples of some gunmetals are CuSn10Zn2 (Admiralty gunmetal) CuSn5Zn5Pb5, and CuSn6-7Zn1.5-2.5Pb0.1-0.5Ni5-5.5. The most widely used cupronickels are the 90/10 and 70/30 types, which generally contain other elements such as iron and manganese to increase resistance to erosion-corrosion, chromium to improve the strength of the alloy and niobium or silicon for casting. For special purposes, for example in seawater containing sand, condenser tubes of the 70/30 type may contain some extra iron and manganese, up to 3%. Due to the important effect of the corrosion-product layer it is very important that it is formed as quickly as possible under conditions which are advantageous for the formation of an even and good-protective layer. Adverse conditions may include sulphides in the seawater which can induce pitting and SCC, and excessive water speeds which may induce local turbulence with subsequent removal of the protective film. As already mentioned, selective corrosion may also be a problem with some alloy types. The seawater quality is very important in relation to the corrosion behaviour, polluted seawater containing sulphides, ammonia or sand being detrimental. Some examples of applications of copper-based alloys are given below:

brasses	- tubesheets, condenser tubing (Al-brass)
bronzes	- pumps, valves, propellers, naval ordnance
Al-bronzes	- pumps, valves, fittings, ship propellers
gunmetals	- pump bodies, valve bodies, water boxes
cupronickels	- condenser tubes, seawater piping

2.7 Titanium

Although in itself a very reactive metal titanium readily forms a very strong passive film with a high ability for rapid self-repair, which makes it very resistant to corrosion. For this reason it is a candidate marine material with excellent corrosion properties in seawater.

It has proved resistant at flow rates as high as 18 m/s. In the presence of sand particles and with flow velocities up to 6 m/s the corrosion rate also remains quite low. Titanium may be alloyed with small percentages of palladium, copper, aluminium and vanadium to improve mechanical properties. The alloys are generally resistant to local corrosion forms such as pitting, crevice corrosion and SCC, although there are some indications that there is a possibility for hydrogen embrittlement. When coupled with other metals titanium generally forms the cathodic (more noble) part, so attention must be paid to bimetallic corrosion of the other part. Stainless steels and aluminium nickel bronzes have been found to suffer only slight attack in seawater when coupled to titanium in a 10:1 anode/cathode ratio. Titanium is often chosen as a replacement material for condenser tubing under seawater conditions where copper alloys have given disappointing results.

2.8 Plastics

Due to their light weight, corrosion resistance and vibration damping qualities, plastics have many potential applications in the marine field. However, these materials lack resistance to higher temperatures and mechanical strength. To improve the mechanical properties they are generally used in composite form. Notwithstanding their excellent resistivity to chemical attack there are some adverse reactions possible such as swelling, ageing and cracking. Well-known composite materials include a polyester resin matrix reinforced with glass fibres. The adhesion between the fibre material and the matrix is important, as is the orientation and average length of the fibres. New developments include vinyl and epoxy resins as matrix materials and Kevlar and graphite fibres as reinforcement. There is interest in further developments aimed at achieving constructions with greater lightness and stiffness. Examples of practical applications are superstructures and hulls for minehunters/sweepers, seawater piping, propellers for ships as well as torpedoes and hatch doors. Apart from application in constructional parts, plastics find also general use as protective coatings on metal surfaces. Both thermosetting and thermoplastic polymers are used for this purpose. Depending on the severity of the environment strengthening with glass fibres or glass flakes can also be envisaged for this kind of application.

2.9 Concrete

Reinforced concrete is used in large quantities for marine applications. Problem areas may be carbonation, which is the chemical degradation of $Ca(OH)_2$ resulting in a decrease of the pH, and the presence of chloride ions. In both cases the problems arise as the carbonation or chloride reaches the steel reinforcement so that diffusion is rate determining. Thus, the best remedy is to apply concrete of high density. This can be accomplished by:

(1) choice of cement type,

(2) amount of cement,

(3) water/cement ratio,

(4) controlled construction, i. e. paying attention to constant conditions during the addition of concrete,

(5) post treatment to oppose cracking, and

(6) extra precautions regarding the reinforcement steel, e. g., coatings or cathodic protection.

2.10 References

General and seawater corrosion data:

1. Corrosion Data Sheets (seawater): W. Katz. Published by Dechema, Frankfurt am Main, FRG, 1976.

2. Sea water Corrosion Handbook: M. Schumacher (ed.). Published by Noyes Data Corp., Park Ridge, NJ, U.S. A.,1979.

3. J.F.G. Condé: in New materials for the marine and offshore industry, Trans I Mar E (TM), 1985, 97, paper 24 .

Materials for seawater handling systems:

4. B. Todd: in Selection of materials for high reliability seawater handling systems; Suppl. to Chemistry and Industry, 2 July, 1977, 14.

5. B. Todd: in Materials for seawater pumps; Proceedings NITO/NIDI conference "Offshore materials and corrosion", Oslo, October 1988.

6. B. Todd and P.A. Lovett: in Selecting materials for sea water systems; Marine Engineering Practice, vol. 1, part 10. Published by The Institute of Marine Engineers/Marine Media Management Ltd.

7. A.H. Tuthill and C.M. Schillmoller: in Guidelines for the selection of marine materials; paper presented at The Ocean Science and Ocean Engineering Conference, organized by the Marine Technology Society, Washington, D.C.,U. S. A., June 14-17, 1965.

8. C. Christensen: in Guidelines for sea water cooling systems covering the corrosion aspect. Report published by Korrosionscentralen, ATV, Copenhagen.

9. British Standard BS MA 18 Aug. 1973: Saltwater piping systems in ships.

10. E.B. Shone: in Problems in Seawater Circulating Systems; Br. Corros. J., 1974, 9,(1), 32-38.

11. R.A. Connell and E.B. Shone: in Seawater Circulating systems; Suppl. to Chemistry and Industry, 2 July, 1977, 23.

12. E.B. Shone and G.C. Grim: in 25 Years Experience with Seawater cooled Heat-transfer equipment in the Shell fleets; Trans. I. Mar. E. (TM), 1986, 98, paper 11.

13. K.D. Efird: in Corrosion within seawater systems on offshore platforms; Proc. Symposium Marine Corrosion on Offshore Structures, organized by Soc. of Chemical Ind. Published by Univ. of Aberdeen, Sept., 1979, 53.

Chapter 3

Specific Cases

PIPES

Piping is used to transport and distribute seawater to heat exchangers, hydrants and other utilities. In this context the term piping encompasses pipes, bends, tees, metering devices, plugs, pockets, strainers and fasteners exposed to seawater.

Compatibility of the materials is of great importance and corrosion due to galvanic action between dissimilar metals is a frequent cause of premature failure, for instance, galvanic corrosion of carbon steel pipes due to nobler metals in valves and pumps or increased sensitivity to corrosion of copper alloys coupled to titanium heat exchangers. The flow velocity in piping may vary with layout of the system, geometry of the actual pipe work and operational utilization of the system. Thus, arduous flow conditions may exist next to, or alternating with, more or less stagnant conditions. Excessively high flow velocities may cause severe corrosion of carbon steel and copper alloys, while stagnant water can be detrimental to stainless steels due to settling of marine fouling and debris.

CASE Pi 1

FLANGE JOINT; MILD STEEL; GALVANIC CORROSION

INSTALLATION: Sea water piping system.

DAMAGE: Leakage at flanges, connected to bronze valves after 3.5 months service.

MATERIAL: Mild steel.

CONDITIONS: The pipes were used to circulate clean aerated sea water at nominal flow velocity of 2 m/s.

CAUSE OF DAMAGE:The corrosion was caused by galvanic action between the mild steel pipe and a bronze valve. Figure 1 shows the severe metal loss and complete wastage of weld metal in the flange directly connected to the valve. For comparison Fig. 2 shows the flange at the opposite end of the same pipe, i.e. 2 metres away from the valve.

REMEDIAL MEASURES: Use compatible materials throughout the system or use coated or lined bronze valves. Use of metallic insulation (breaking the electron path) may also be considered, but external short circuiting through the support structure must also be avoided.

COMMENTS: Mild steel even without may galvanic effects, may suffer enough corrosion to make it an unattractive material for seawater piping.

Figure 1: Weld metal wasted (note screwdriver) and pipe wall thinned by galvanic corrosion.

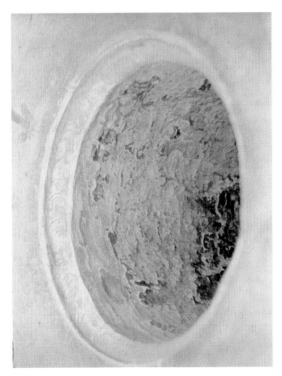

Figure 2: Flange end 2 m from bronze valve. Note the much lower corrosion in this flange compared to Fig. 1.

CASE Pi 2

PIPING; GALVANIZED MILD STEEL; LOCALIZED CORROSION

INSTALLATION: Seawater cooling system on a coaster type ship.

DAMAGE: Local deep corrosion attack in 3 inch diameter cooling water pipe, Figs. 1 and 2.

MATERIAL: Hot-dip galvanized steel.

CONDITIONS: Part time stagnant seawater at 45°C.

CAUSE OF DAMAGE: The zinc layer is designed to corrode sacrificially to protect any bare steel as long as some zinc remains. However, it may become passive if exposed to stagnant polluted sea water due to the formation of zinc sulphide, so preventing sacrificial corrosion. As the galvanized layer contained porosity corrosion of the steel substrate occurred leading to a pitting type of attack.

REMEDIAL MEASURE: Avoid warm stagnant seawater, by draining or by maintaining a sufficiently high flow in the system.

Copper alloys should not be used as a substitute in these circumstances as they will be prone to corrosion in subsequent service if a sulphide film is allowed to form.

Titanium, high molybdenum (> 6%, Mo) stainless steel, plastics or glass fibre reinforced polyester can give excellent service.

Figure 1: View of pipe interior.

Figure 2: As above after cleaning.

CASE Pi 3

PIPE BEND; COATED MILD STEEL; CORROSION

INSTALLATION: Seawater cooling system on a ship.

DAMAGE: Corrosion of the pipe wall has occurred after 3 months of service (see Fig.).

MATERIAL: The piping system was made from epoxy coated mild steel.

CONDITIONS: The piping carried clean seawater.

CAUSE OF DAMAGE: The corrosion of the pipe wall was caused by seawater gaining access to its surface at defects in the coating. This may have happened as a result of poor preparation of the metal prior to coating and/or poor application of the coating.

REMEDIAL MEASURES: The affected area should be cleaned and, following adequate surface preparation, the coating repair can be undertaken.

COMMENTS: Failure of protective coatings do occur quite frequently but usually only small areas are affected and local repair is sufficient.

The main cause of serious coating failure is usually poor surface preparation of the metal prior to coating. In the case of piping it is important to carry out coating inspection regularly as any breakdown can result in rapid metal loss due to the presence of more noble metals (tube plates and tubes) in the system.

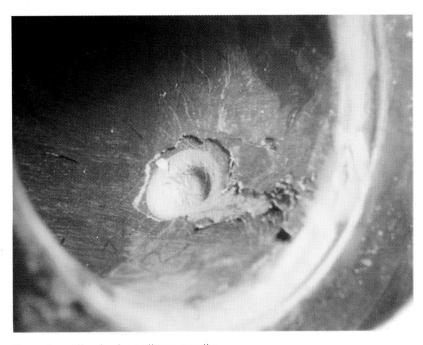

Corrosion attack at coating porosity.

CASE Pi 4

PIPE JOINT; GREY CAST IRON; GRAPHITIC CORROSION

INSTALLATION: Pipe joint in power station cooling water system.

DAMAGE: Loss of strength due to graphitic corrosion.

MATERIAL: Grey cast iron.

CONDITIONS: Seawater with higher than normal suspended solids content.

The system had given very good service, i.e. in excess of 20 years. An inspection of the cooling water system was carried out in order to assess the need for refurbishment.

CAUSE OF DAMAGE: It was found that enchanced attack, taking the form of graphitic corrosion, had occurred at pipe joints of the "Dresser" type coupling.

Figure 1 shows a section through the pipe wall near the end of the pipe and significant metal loss with graphitic residue can be seen on both the external and internal surface. Figure 2 shows a section through the pipe end and through-wall graphitic corrosion can be seen at its edge. The more severe corrosion at the external surface of the pipe end, within the coupling, may be caused by slight galvanic effect from the coupling material or the development of low pH in the stagnant area on the external pipe surface.

COMMENTS: This type of graphitic corrosion is typical of grey cast iron exposed to seawater. Due to heavy wall thicknesses grey cast iron may, however, give satisfactory service; in this case more than 20 years.

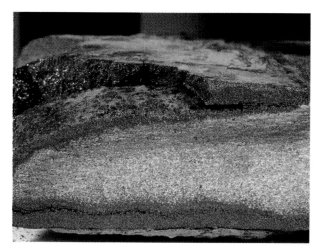

Figure 1: Cross section of pipe end.

Figure 2: As Fig. 1, showing more corrosion at the outer surface exposed to stagnant conditions within the coupling area.

CASE Pi 5

PIPE BEND; GREY CAST IRON; GRAPHITIC CORROSION

INSTALLATION: Pipe bend in power station cooling water system.

MATERIAL: Grey cast iron.

DAMAGE: Severe metal loss in pipe bend.

CONDITIONS: Brackish water with very high suspended solids content.

The system had given very good service, i.e. in excess of 20 years. An inspection was carried out in order to assess the need for refurbishment of the system.

CAUSE OF DAMAGE: The severe metal loss was confined to parts of the system experiencing higher than average cooling water flows such as bends (see Fig.). In these areas, oxygen is readily available and this may sustain the graphitic corrosion.

One important point to note is that the extent of metal loss of cast iron components is very difficult to assess visually due to the nature of graphitic corrosion.

The pipe bend in question was thoroughly shot blasted prior to inspection.

COMMENTS: Graphitic corrosion in grey cast iron is typical in seawater service. The rate of attack usually diminishes with time as the graphitic layer increases in thickness and acts as a barrier to the outward diffusion of corrosion products. If oxygen is readily available at the graphitized surface, the corrosion process may however continue. If the flow conditions are such that the graphitized layer is removed by hydrodynamic forces, corrosion may be rapid.

If the graphitized layer is damaged on purpose, e. g., to estimate corrosion depth, it is essential to protect the bare metal by coating or cathodic protection. Otherwise local penetrations of the wall may be extremely rapid.

Extensive wastage of material from inner surface.

CASE Pi 6

PIPE NOZZLE; CuNi10Fe; GALVANIC CORROSION

INSTALLATION: Generator cooler on a dredging vessel.

DAMAGE: Perforation of cooling water piping at cooler outlet (see Fig.).

MATERIAL: Piping: CuNi10Fe
               Plate cooler: Titanium

CONDITIONS: Cooling water: seawater at a flow rate of 0.2-1.5 m/s. The system had been used for approximately 8 weeks.

CAUSE OF DAMAGE: At the cooler outlet a CuNi10Fe reducer and elbow were directly connected to the cooler body. The inlet piping was connected by means of rubber bellows. From the outlet piping both elbow and reducer showed severe attack and perforation. The difference in potential between CuNi10Fe and titanium in seawater is considered to be the primary cause of failure.

REMEDIAL MEASURES: Electrical separation of the piping and the cooler, preferably followed by a resistance measurement to check this isolation.

Perforation of pipe wall at arrow.

CASE Pi 7

PIPE BRANCH; CuNi-ALLOY; EROSION-CORROSION

INSTALLATION: Seawater cooling system.

DAMAGE: Excessive wall thinning in pipe downstream of branching (see Fig.).

MATERIAL: Copper nickel alloy.

CONDITIONS: Flowing seawater.

CAUSE OF DAMAGE: The average design flow velocity may have been satisfactorily low but the arduous flow condition caused by the abrupt change of flow direction has caused erosion corrosion.

REMEDIAL MEASURE: Damage of this type can be prevented by improving the layout of the piping system and maintaining a proper hydraulic balance in the system. Sharp tees, as in this case, should be avoided, for example, by using saddle type branching.

View of inner pipe surface immediately down stream of branch.

CASE Pi 8

PIPE; CuNi10Fe1.5; EROSION-CORROSION

INSTALLATION: Seawater piping system.

DAMAGE: Pitting and perforation of the wall of a pipe, Figs. 1, 2 and 3.

MATERIAL: Copper nickel piping meeting the requirements of CuNi10Fe1.5.

CONDITIONS: The average design flow velocity in the piping system was 2 m/s. A butterfly valve was situated directly before the bend which had a radius of 200 mm. The inner diameter of the pipe was 120 mm.

CAUSE OF DAMAGE: The butterfly valve in the piping system had caused excessive local turbulence downstream of it. This turbulence had caused impingement attack to occur, the damage pattern following the screw-like motion of the water (Fig. 2). The most severe attack and perforation had occured at a weld in the pipe (Fig. 3).

REMEDIAL MEASURES: Damage of this type can be prevented by improving the layout of the piping system. As a general rule an allowance of 5 times the diameter of the pipe should be made between turbulence raisers and bends. Hence a valve should never be positioned directly before a bend. If design changes cannot be tolerated then the use of alternative more erosion/corrosion resistant materials should be considered.

Figure 1: Corroded pipe bend.

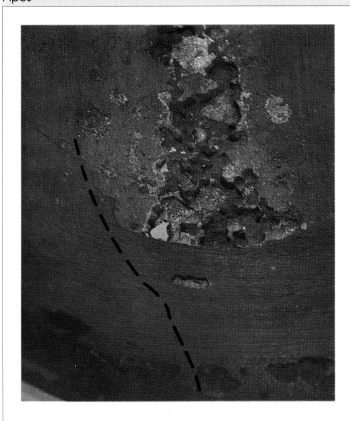

Figure 2: The erosion-corrosion attack follows a screw-line as shown by the dashed line.

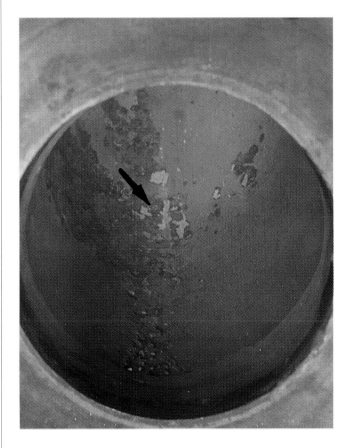

Figure 3: Perforation of pipe wall at arrow.

CASE Pi 9

PIPE; CuNi10Fe; EROSION-CORROSION

INSTALLATION: Seawater piping system.

DAMAGE: Pitting of the type shown in Fig.1. The damage was most severe in areas adjacent to protrusions such as gaskets between flanges, welds and bends.

MATERIAL: Copper nickel piping meeting the requirements of CuNi10Fe.

CONDITIONS: The seawater velocity in the pipework was in excess of 4 m/s.

CAUSE OF DAMAGE: The pitting damage was of the type associated with impingement attack (erosion-corrosion) and had been caused by the excessive velocity of the seawater. Additionally protrusions within the pipework would have acted as turbulence raisers and in areas of high water turbulence the corrosion/erosion damage was increased as Fig. 2 shows.

REMEDIAL MEASURES:The flow velocity in the system was restricted to 2 m/s by installing a pump of the appropriate capacity. This water velocity is lower than the maximum generally recommended for this alloy. This lower water velocity was suggested since it was not practical to remove the protrusions from within the pipes and hence the turbulence associated with them. If no protrusions were present in the pipes water velocities of 3 m/s could have been used.

If the water velocities in the original system (4 m/s) had really been necessary, then it would have had to be redesigned, either using piping of larger diameter or alternatively more erosion-corrosion resistant materials.

Figure 1: Severe pitting-like attack downstream of protruding gasket at flange joint.

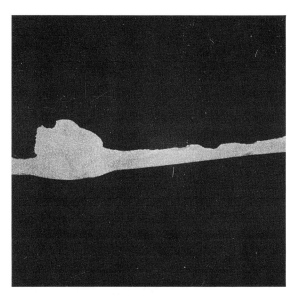

Figure 2: Excessive wall thinning due to erosion-corrosion caused by protruding weld bead.

CASE Pi 10

FLANGE JOINT; CuNi5Fe; EROSION-CORROSION

INSTALLATION: Seawater piping system.

DAMAGE:Perforation of the pipe behind a flange connection (Figs. 1 and 2).

MATERIAL:Pipe and flanges meeting the requirements of CuNi5Fe.

CONDITIONS: Flowing seawater of ambient temperature; designed average flow velocity 1.5 m/s.

CAUSE OF DAMAGE: Erosion-corrosion due to locally high turbulence associated with misalignment of the outlet between the flanges. The same kind of damage can occur when the tubes are misaligned.

REMEDIAL MEASURES: Better control of alignment of gasket and tube flanges is required to prevent damage of this type. The choice of CuNi5Fe for this purpose is questionable since this alloy is susceptible to erosion corrosion; Al-brass, CuNi10Fe or CuNi30Fe would have been more suitable materials of construction, since they are all more resistant to erosion- corrosion damage.

COMMENTS: Whatever material is used, the presence of turbulence raisers in a seawater piping system should be avoided.

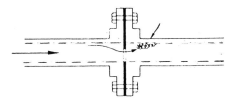

Figure 1: Perforation of pipe wall downstream from protruding gasket.

Figure 2: Inner surface of pipe section in Fig.1. Excessive wall thinning and perforation.

CASE Pi 11

FLANGE JOINT; AUSTENITIC STAINLESS STEEL; CREVICE CORROSION

INSTALLATION: Piping system handling water from the Mediterranean Sea.

DAMAGE: Severe corrosion of the type seen in the Fig. occurred beneath the gaskets of all flanges.

MATERIAL: Pipes and flanges of stainless steel meeting the requirements of AISI 321.

CONDITIONS: The damage was observed after the piping system had been in intermittent service with clean filtered seawater for 16 months.

CAUSE OF DAMAGE: The piping system had been manufactured from a material which is prone to crevice corrosion in seawater.

REMEDIAL MEASURES: It is possible to alleviate such corrosion by applying gaskets with in-built cathodic protection, e.g. lead or zinc. The use of petrolatum products with zinc-oxide filler, as a gasket compound, can be used as a temporary protection against crevice corrosion. The problem may also be solved by the use of alternative more crevice corrosion resistant materials for such applications.

Crevice corrosion at flange face. Stainless steel type AISI 321.

CASE Pi 12

PIPING; AUSTENITIC STAINLESS STEEL; CREVICE CORROSION

INSTALLATION: Piping system carrying seawater.

DAMAGE:Leakage occurred at the flanges after about 2 months in service. Upon detailed examination it was found that the flanges were heavily pitted and that pipes were also pitted in a random manner.

MATERIAL: Austenitic stainless steel to AISI 316.

CONDITIONS: The piping system contained seawater at 10-15° C flowing at about 2.0 m/s.

CAUSE OF DAMAGE: Crevice corrosion had occured within the flanges (see Fig.) and pitting corrosion on the pipe walls. This alloy is very susceptible to these forms of corrosion and is unsuitable for use in seawater handling systems.

REMEDIAL MEASURES: Use an alternative alloy such as high molybdenum (> 6% Mo) stainless steel or the 90/10 copper nickel alloy. These alloys are considerably more resistant to crevice and pitting corrosion. If the envisaged water velocity is high the copper nickel alloy may be prone to erosion-corrosion damage. Cathodic protection can be used to alleviate pitting in crevice corrosion damage on AISI 316 type stainless steel. See also CASE Pi 11.

Crevice corrosion at flange face.

CASE Pi 13

FLANGE JOINT; DUPLEX STAINLESS STEEL; CREVICE CORROSION

INSTALLATION: Piping system carrying seawater.

DAMAGE: Leakage had occurred at the flanges after about 6 months in service. Upon detailed examination it was found that the flange faces were severely attacked (see Fig.).

MATERIAL: Duplex stainless steel to UNS S31803.

CONDITIONS: The piping system contained seawater at 10-15° C flowing at 2 m/s.

CAUSE OF DAMAGE: Crevice corrosion had occurred within the flanges. This alloy is susceptible to crevice corrosion and as such is unsuitable for use in seawater handling systems.

REMEDIAL MEASURES: Use an alternative alloy such as high molybdenum (> 6% Mo) stainless steel or the 90/10 copper nickel alloy. These alloys are considerably more resistant to crevice corrosion.

If the expected water velocity is high the copper nickel alloy may be prone to erosion-corrosion damage. Cathodic protection can be used to alleviate pitting or crevice corrosion damage on UNS S31803 type stainless steels. See also CASE Pi 11.

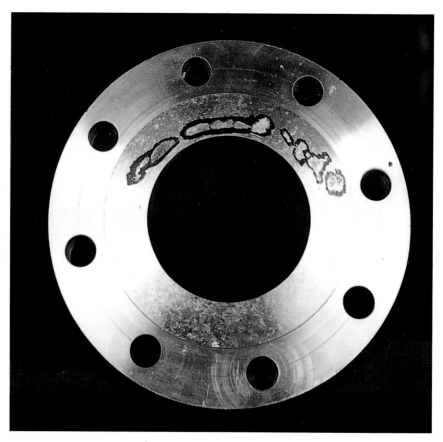

Crevice corrosion at flange face.

CASE Pi 14

PIPING; AUSTENITIC STAINLESS STEEL; PITTING

INSTALLATION: Seawater fire fighting and utility system.

DAMAGE: Leaking pipes due to local corrosion (pitting) along girth welds (see Fig.).

MATERIAL: Pipes made of AISI 316 L.

CONDITIONS: The pipes were carrying clean cold seawater yet most of the time the system was idle and partly empty.

CAUSE OF DAMAGE: Pitting corrosion in weld zone. The corrosion resistance of weld zones is less than that of the base metal especially in the heat tinted zone arising from welding with insufficient inert gas cover.

REMEDIAL MEASURES: While AISI 316 L may be usable in cold clean seawater service (<10°C) it does require welding with sufficient inert gas root side cover to avoid loss of corrosion resistance.

COMMENTS: Stainless steel type AISI 316 is generally not recommended for seawater applications. Use the high molybdenum (> 6% Mo) stainless steel, CuNi alloys, plastics or glass fibre reinforced polyester.

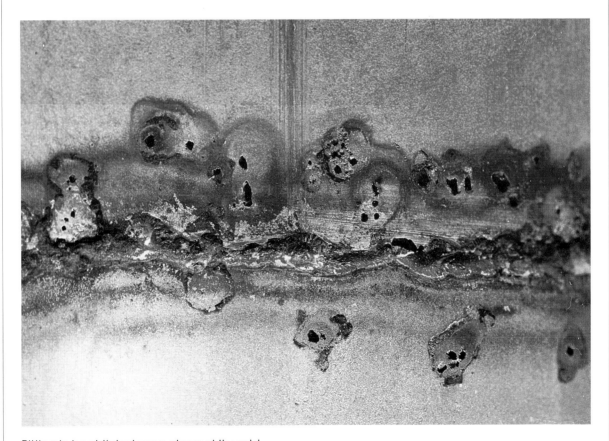

Pitting in heat tinted zone along girth weld.

CASE Pi 15

PLUG; FERRITIC STAINLESS STEEL; PITTING

INSTALLATION: Plug in a seawater piping system.

DAMAGE: Pitting attack of plug (see Fig.).

MATERIAL: Ferritic stainless steel type X8Crl7 plug in a AISI 316 type austenitic stainless steel pipe.

CONDITIONS: Seawater of ambient temperature, average flow velocity 2 m/s, also prolonged periods of stagnation.

CAUSE OF DAMAGE: The type of steel used for the manufacture of the plug is not resistant to corrosion in seawater; the intensity of the attack is increased by galvanic contact with more noble stainless steel pipework.

REMEDIAL MEASURE: A plug of material similar in composition or more noble than that of the pipework should be used. However, AISI 316 is not suitable for use in seawater unless it is protected by a cathodic protection system.

Heavily pitted plug, ferritic stainless steel.

CASE Pi 16

PLUG; BRASS; DEZINCIFICATION

INSTALLATION: Seawater piping system.

DAMAGE: Dezincification of deaeration plug, Figs. 1 and 2.

MATERIAL: Plug: alpha-beta brass
Pipes: CuNi10Fe.

CONDITIONS: Flowing seawater of ambient temperature, flow velocity 1.5 m/s; idle periods.

CAUSE OF DAMAGE: Alpha-beta brass is prone to corrosion in seawater. The attack takes the form of selective leaching of the zinc constituent (dezincification) leaving (redeposited) copper to maintain the original shape. The dezincification attack in this case is possibly aided by galvanic difference between plug and piping.

REMEDIAL MEASURES: Plugs to be made of the same alloy as the piping itself.

Figure 1: Cross section of brass plug showing severely dezincified zone at arrow.

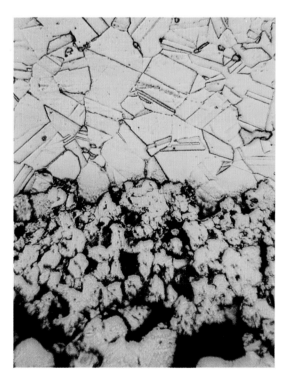

Figure 2: Microstructure of dezincification zone.

CASE Pi 17

**PIPE BEND; ALUMINIUM BRASS; STRESS
CORROSION CRACKING**

INSTALLATION: Seawater cooling system on
board a naval vessel.

DAMAGE: Repetitive leakages due to cracking
(see Fig.).

MATERIAL: Aluminium brass, to BS2871.

CONDITIONS: Flowing seawater at ambient
temperature.

CAUSE OF DAMAGE: Cold bending of pipe
leaving sufficient residual tensile stresses to
cause stress corrosion cracking of this
particular alloy. Attemps to repair leakages by
brazing often resulted in more cracking.

REMEDIAL MEASURES: Stress relieving heat
treatment after cold forming following the
suppliers instruction or use of hot bending.

Pipe bend showing cracks and attempted repair of cracks by
brazing.

CASE Pi 18

MANOMETER SOCKET; ALUMINIUM BRASS; FATIGUE CRACKING

INSTALLATION: Seawater cooling system on a ship.

DAMAGE: Crack and leak in fillet of soldered on socket, Figs. 1 and 2.

MATERIAL: Socket: naval brass
Solder: lead-tin alloy
Main pipe: aluminium brass

CONDITIONS: Ambient temperature in machinery room.

CAUSE OF DAMAGE: The overhanging manometer was excited by structural vibrations (always present on ships). The lead-tin solder has a very low fatigue strength and once the fatigue crack had formed it progressed through the metallurgical bond into the substrate metal.

REMEDIAL MEASURE: Support the manometer properly or use welded-on sockets that have a higher fatigue endurance limit.

Figure 2: Cross section showing fatigue crack starting in the soft solder.

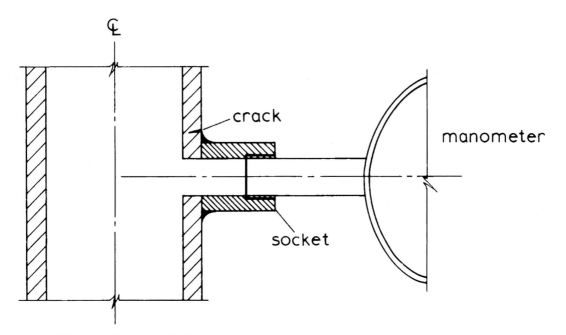

Figure 1: Schematic presentation of damage.

CASE Pi 19

THERMOMETER POCKET; BRASS; DEZINCIFICATION

INSTALLATION: Seawater cooling system on a bulk carrier.

DAMAGE: Dezincification and pitting (see Fig.).

MATERIALS: Brass (63Cu-37 Zn).

CONDITIONS: Seawater flow velocity 0-2.5 m/s, temperature < 32°C.

CAUSE OF DAMAGE: This type of brass is sensitive to dezincification in seawater. The corrosion rate was increased galvanically by contact with the more noble CuNi 90/10.

REMEDIAL MEASURES: Use of more resistant materials like CuNi 90/10 or tin bronze.

COMMENTS: This type of brass should not be used in seawater.

Galvanically stimulated attack of brass (63Cu-37Zn) thermometer jacket.

CASE Pi 20

FLOW METER; FREE MACHINING AUSTENITIC STAINLESS STEEL; CREVICE CORROSION

INSTALLATION: Meter for measuring flow velocity.

DAMAGE: Crevice corrosion of housing and nut (see Fig.).

MATERIALS: Stainless steel AISI 303.

CONDITIONS: The meter was in intermittent service in flowing seawater at ambient temperature during a 3 month period.

CAUSE OF DAMAGE: Corrosion had occurred in crevices and beneath marine fouling.

Whilst the attack was due to crevice corrosion it seems probable that, in this case, it had been enhanced by the presence of non-metallic inclusions in the AISI 303 steel.

REMEDIAL MEASURES: Alternative materials that are more crevice corrosion resistant should have been used (e.g. CuNi 90/10, CuNi 70/30, different types of bronzes, Monel, high alloy stainless steels). Cathodic protection is another possibility.

COMMENTS: This type of attack is typical for non-protected conventional stainless steels in seawater when crevices are present. See also CASE Pi 11.

Crevice corrosion of AISI 303 stainless steel flow meter.

CASE Pi 21

RETAINER BOLT; AUSTENITIC STAINLESS STEEL; CREVICE CORROSION

INSTALLATION: Seawater exposure rig.

DAMAGE: Crevice corrosion of threaded area of bolt beneath nut (see Fig.).

MATERIALS: AISI 316 austenitic stainless steel nuts and bolts were used to secure pieces to an epoxy coated carbon steel rig.

CONDITIONS: The rig was exposed to seawater of ambient temperature at a depth of 45 m, for about one year.

CAUSE OF DAMAGE: Crevice corrosion of screw thread had occurred. Generally austenitic stainless steels of the 304 and 316 series are susceptible to crevice corrosion unless additional protective measures are taken. In this case cathodic protection could have been obtained from the carbon steel rig, but the epoxy coating eliminated this possibility.

REMEDIAL MEASURES: Cathodic protection e.g. from non-coated carbon steel. Use of more crevice corrosion resistant materials is another possibility.

COMMENTS: This type of attack is typical for non-protected conventional stainless steels in seawater when crevices are present. See also CASE Pi 20.

Crevice corrosion of AISI 316 stainless steel bolt and nut.

CASE Pi 22

STRAINER; BRASS; DEZINCIFICATION

INSTALLATION: Seawater piping system aboard ship.

DAMAGE: Broken inlet strainer (Fig. 1).

MATERIAL: Brass containing 40% zinc in galvanic contact with a gunmetal housing.

CONDITIONS: The inlet strainer is used to protect the sea water pump supplying the cooling system with fresh sea water of ambient temperature against debris, etc.; the flow velocity was < 2 m/s.

CAUSE OF DAMAGE: The 40% zinc containing brass is a two phase alloy, consisting of a copper-rich alpha- and a zinc-rich beta-phase (Fig. 2, p. 50). In particular the beta-phase is sensitive to dezincification, the copper remaining as a porous mass without any mechanical strength (Figs. 3 and 4, p.50).

REMEDIAL MEASURE: (1) the use of a material which is less sensitive to the dealloying form of local corrosion, as for instance gunmetal, (2) the use of carbon steel protected with an appropiate coating (paint, PVC, PE) preferably in combination with galvanic anodes.

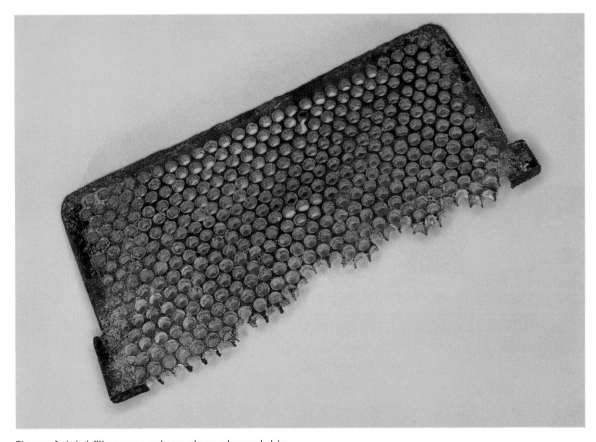

Figure 1: Inlet filter sea water system aboard ship.

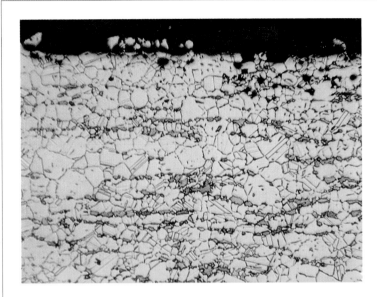

Figure 2: Microphotograph of dezincified and underlying zone, showing alpha and beta crystals. Magnification 75x.

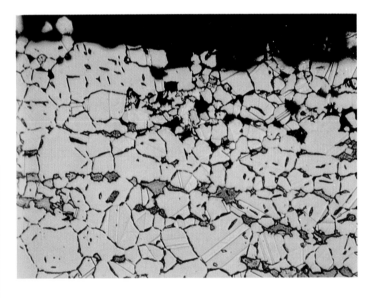

Figure 3: Detail of Figure 2. Magnification 150x.

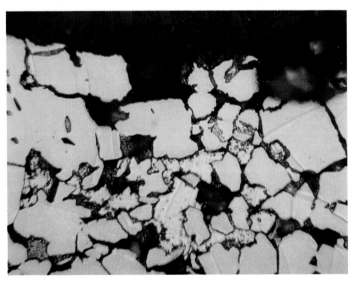

Figure 4: Detail of Figure 2. Magnification 350x.

PUMPS

The function of a pump in a seawater system is to convert power from a prime mover to an impeller which causes flow in the seawater. This flow can be converted to pressure by means of a casing arranged around the impeller.

Centrifugal pumps are the type most commonly used and the main components are the impeller, casing and shaft.

Examples of corrosion commonly experienced in pump components are provided in the following cases. Two examples relate to a ships propeller and propeller shaft. These are included as they have some features in common with pump components.

Stainless steels are often used in pumps for components such as shafts and fasteners. Also, because of their very high resistance to turbulent flow in seawater they are often chosen for impellers. Examples of pitting, crevice corrosion and corrosion-fatigue in stainless steel pump parts are given.

Copper-base alloys can suffer corrosion-erosion in flowing seawater and examples of this are provided for both pump body and impeller. Similar corrosion behaviour is shown by Ni-Resist alloy cast iron and ordinary grey iron and examples are given.

Cavitation damage can occur in pump parts and examples of this on a nickel-copper alloy and on grey cast iron are provided.

Grey cast iron can suffer graphitisation - a form of selective corrosion - which produces a corrosion product with a high graphite content. One example of this is described.

A stress corrosion cracking failure in a manganese bronze propeller from a ship is included as this type of failure can occur in pump components.

CASE Pu 1

PUMP SHAFT; TYPE 431 STAINLESS STEEL; PITTING

INSTALLATION: Pump for seawater supply to cooling system.

DAMAGE: The shaft of the pump was attacked in the form of pits (Figs. 1 and 2); a locking screw attached to the shaft was heavily corroded, as was the corresponding screw thread on the shaft.

MATERIAL: Shaft: stainless steel (AISI 431 DIN: X22CrNi17)
Pump case ⎫
Propeller ⎬ Bronze (CuSn14)
Locking Nut ⎭

CONDITIONS: Ambient temperature seawater.

CAUSE OF DAMAGE: Type 431 stainless steel is prone to pitting in seawater. Galvanic coupling of the stainless steel shaft to the copper alloy parts of the pump, raised the potential of the shaft to such a level as to provoke pitting as well as crevice corrosion in the crevice which was formed between the locking nut and the shaft. The locking nut itself was found to have deteriorated with a red discoloration which is a sign of dezincification (Fig. 3). The locking nut was not fabricated as specified: instead of bronze an inferior seawater resistant material, brass had been used.

REMEDIAL MEASURES: The application of better corrosion resistant materials for the parts of the pump, for instance:
Shaft   Monel K-500

Locking nut ⎫
Propeller   ⎬ CuAl10Ni (DIN 1714)

COMMENT: Stainless steels such as Type 431 have poor resistance to pitting and crevice corrosion in seawater but are often chosen because they are inexpensive and have attractive mechanical properties. Materials with better resistance to seawater corrosion are necessary for pump shafts.

Figure 1: General appearance of corroded pump shaft.

Figure 2: Pitted area on Type 431 stainless steel pump shaft.

Figure 3: Section of brass locking nut showing dezincifation.

CASE Pu2

PUMP SHAFT; TYPE 321 STAINLESS STEEL; FATIGUE

INSTALLATION: Ship's seawater/cargo pump.

DAMAGE: Fracture of pump shaft.

MATERIAL: Free machining stainless steel of AISI 321 type.

CONDITIONS: Intermittent pumping of seawater or crude oil.

CAUSE OF DAMAGE: The 108mm diameter shaft had fractured due to fatigue originating from several sites in a circumferential groove located between the seals and the main bearing. The fatigue crack had propagated transversly across about half of the shaft and the final fracture occurred by a ductile shearing mechanism (see Fig.). The stresses required for this to occur had arisen from high torsional loadings and possibly a misaligned shaft.

The shaft material was austenitic with a relatively high concentration of longitudinal banded inclusions. The cracks were transgranular, typical of those associated with fatigue, but in the final fracture zone the gross plastic deformation at the surface indicated that the final fracture had occurred by ductile shear. The fracture face was coated with rust. The Vickers hardness of the shaft was about 160HV30.

REMEDIAL MEASURES: Inspect the assembly for signs of misalignment and replace the shaft with a stronger alloy that has a corrosion resistance at least that of AISI316 such as a duplex stainless steel or Monel K500.

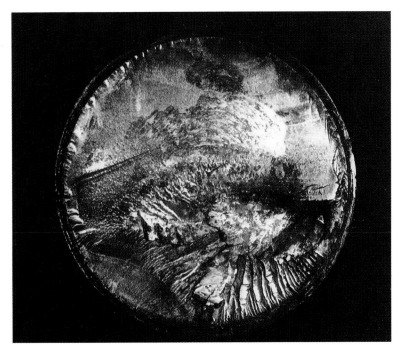

Fracture surface of Type 321 pump shaft.

CASE Pu3

PUMP SHAFT; TYPE 431 STAINLESS STEEL; CREVICE CORROSION

INSTALLATION: Seawater pump.

DAMAGE: Corroded pump shaft (see Fig.).

MATERIALS: Pump shaft: martensitic stainless steel AISI 431
Impeller: rubber lined bronze

CONDITIONS: Seawater at ambient temperature, prolonged idle periods.

CAUSE OF DAMAGE: Crevice corrosion between pump shaft and impeller initiated during idle period. Due to the stagnant conditions, deaeration occured inside the crevice, ultimately leading to crevice attack.

REMEDIAL MEASURES: 1. Alternative materials. 2. Remove the seawater, rinse with potable water and dry when not in use for prolonged periods.

COMMENT: See CASE Pu1.

Crevice corrosion of Type 431 stainless steel shaft.

CASE Pu4

PUMP SHAFT; DUPLEX STAINLESS STEEL; CREVICE CORROSION

INSTALLATION: Seawater pump in a chemical plant cooling water system.

DAMAGE: The shaft had suffered crevice corrosion beneath the rubber liner of co-rotating parts of the mechanical seal after one year of service. Ultimately the mechanical seal leaked beneath the rotating element (see Fig.).

CONDITIONS: Seawater.

MATERIALS: Shaft - AISI 329 stainless steel
Pump housing and impeller - bronze.

CAUSE OF FAILURE: This type of stainless steel is prone to crevice corrosion in seawater, especially warm seawater.

REMEDIAL MEASURES: Use a more resistant stainless steel. Use a corrosion resistant lining on the shaft beneath the mechanical seal. Bronze or plastics liners are satisfactory but only if the crevice between the liner and shaft is avoided by design or use of a crevice-filling cement.

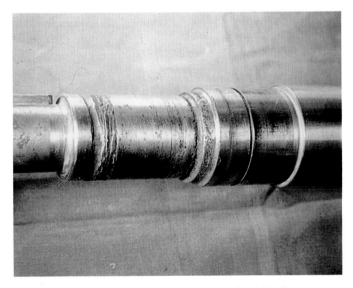

Crevice corrosion of Type 329 stainless steel shaft.

CASE Pu5

PUMP BODY; GUNMETAL; EROSION-CORROSION

INSTALLATION: Ship's seawater circulating pump - Fig. 1.

DAMAGE: Erosion-corrosion of a high speed centrifugal pump body had occurred. Figure 2 shows the severe impingement attack which had caused perforation of the wall. The damage was most pronounced in the discharge area.

MATERIAL: The pump body was manufactured from leaded gunmetal of the BS1400 : 1961-LG2-C type.

CONDITIONS: This pump had been fitted into a non-ferrous seawater circulating system and perforation had occurred after only two years of operation.

CAUSE OF DAMAGE: The pump casing had suffered severe impingement attack which was associated with high velocity turbulent sea water at the impeller discharge. This damage may have been intensified by air entrained in the seawater. It is considered that a leaded gunmetal of the type used is unsuitable for this application.

REMEDIAL MEASURES: The application of a more corrosion resistant material such as a nickel aluminium or a chromium hardened cast copper nickel is recommended. Certain of the high quality stainless steels may also be suitable. However, it must be remembered that the pump body is only part of a system and whatever material selected must be galvanically compatible with the internals of the pump and the rest of the system.

Figure 2: Erosion-corrosion of gunmetal pump casing.

Figure 1: Perforation in gunmetal pump casing.

CASE Pu6

PUMP IMPELLER; TIN BRONZE; EROSION-CORROSION

INSTALLATION: Impeller from a seawater circulating pump in a closed system supplying a water-jet ejector to extract gases from the condenser of a thermal power plant.

MATERIAL: Tin bronze ISO - CuSn12, ASTM C90800 (Cu-12Sn).

CONDITIONS: Seawater, temperature 20 - 40°C and free from sulphides or air bubbles.

DAMAGE: Heavy erosion-corrosion and pitting, mainly at the leading edge of the impeller after only 12 months service. Impeller diameter 330mm and rotating speed 1750 rpm. Design of the pump inlet and the inlet pressures were correct (see Fig.).

REMEDIAL ACTION: Attempts to reduce corrosion by applying cathodic protection with steel pipes connected direct to the pump were unsuccessful. The bronze pumps were later replaced by stainless steel pumps (DIN W. No. 1.4450-X8 Cr Ni Mo 275 - (Fe - 27Cr - 4.5Ni - 1.5Mo).

COMMENT: Copper-base alloys are prone to rapid erosion-corrosion in seawater if flow velocities for the alloy are exceeded - as in this case. When a copper-base alloy is chosen for a pump impeller, it is necessary to ensure that flow velocities are acceptable for the alloy.

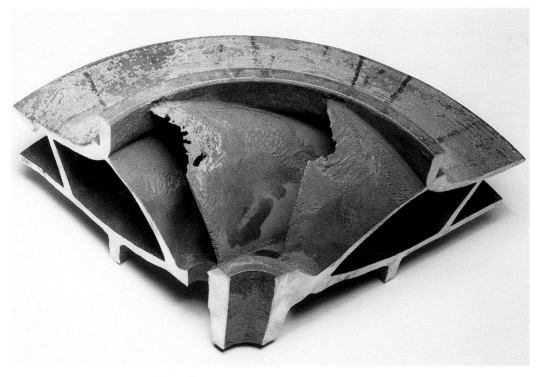

Erosion-corrosion of pump impeller.

CASE Pu7

PUMP IMPELLER; NICKEL-COPPER ALLOY; CAVITATION

INSTALLATION: Ship's seawater circulatory pump.

DAMAGE: The leading edge of the suction sides of the impeller blades had been subjected to severe metal wastage of the type that could be attributed to cavitation erosion. The damage became apparent after about 4 years in service.

MATERIAL: The impeller was of the Monel type although as the analysis showed, it did not meet the compositional requirements laid down in BS3071:1959.

CONDITIONS: The impeller had been fitted into a high speed centrifugal pump made from a leaded gunmetal. This pump was part of a non-ferrous system that handled clean seawater on an intermittent basis.

CAUSE OF DAMAGE: A detailed examination showed that the impeller was very porous and cracking was observed. It had been subjected to a form of selective attack termed denickelification. This denickelification process is thought to be associated with the very poor quality of the casting and the conditions which occur in pumps when they are not in use and contain stagnant sea water.

The metal wastage at the leading edges of the suction sides of the impeller blades is attributed to cavitation erosion. However, the absence of work hardening in the eroded areas suggests that the intensity of cavitation was low. Nevertheless, this would probably have been sufficient to cause the wastage observed because of the corroded nature of the material.

REMEDIAL MEASURES: Alternative materials of construction for impellers should be considered. Nonmetallic materials such as the filled phenolic resins and nylons have been found to give satisfactory in-service performance as have some of the high performance stainless steels.

Figure 1: Cavitation damage on nickel-copper alloy impeller.

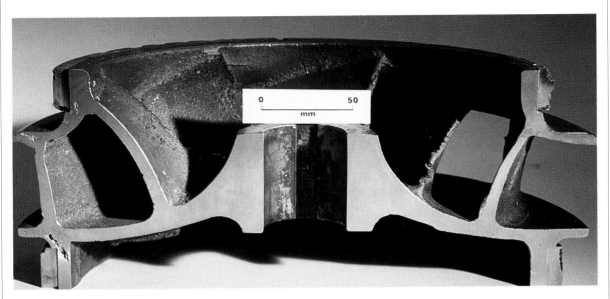

Figure 2: Section of impeller.

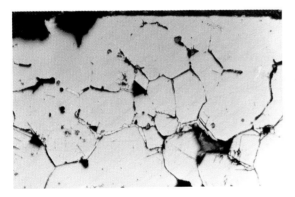

Figure 3: Microstructure of impeller material showing denickelification in the grain boundaries.

CASE Pu8

PUMP IMPELLER; CAST IRON; EROSION-CORROSION AND GRAPHITISATION

INSTALLATION: The impeller was fitted to a fire system pump on an offshore platform.

DAMAGE: General graphitization of the impeller surface to a depth of 1.5-2mm with local deep corrosion attack had occurred after 2 years service. This had led to perforation at the vane inlet edge (see Fig.).

CONDITIONS: Seawater.

MATERIAL: Grey cast iron.

CAUSE OF FAILURE: Cast iron is not resistant to corrosion in seawater unless the corrosion product is retained in the matrix of graphite left by corrosion. The retained layer takes the same form as the original metal but has virtually no tensile strength. In areas where hydrodynamic and/or mechanical forces act, the layer may be removed and the process repeated over and over again.

As this pump had been in use only for very few hours in total (during trials and fire drills) as contrasted to a total immersion time of 2 years, the poor corrosion resistance offered by the cast iron impeller shows clearly that such a pump could not be trusted during a prolonged period of use.

REMEDIAL MEASURES: Install pumps made of materials with improved seawater corrosion resistance.

COMMENT: Corrosion of unalloyed cast iron in seawater increases rapidly with flow velocity. Cast iron is unsuitable for use in seawater where flow velocity is high as in a pump impeller.

Erosion-corrosion of cast iron impeller.

CASE Pu9

PUMP IMPELLER; CAST IRON; CAVITATION

INSTALLATION: Sugar plant, waste water pump handling.

DAMAGE: The blades of the impeller suffered severe cavitation damage showing the characteristic sponge-like appearance on the suction side. The pump had been in service for only 6 weeks (see Fig.).

CONDITIONS: Waste water from a sugar plant - seawater.

MATERIALS: Cast iron.

CAUSE OF FAILURE: Flow conditions in a pump can result in the pressure falling below the vapour pressure of the seawater, so that vapour cavities form. When these bubbles are subsequently swept into areas of higher pressure they may collapse, causing high velocity water jets to impinge on the metal surface. The repeated formation and collapse of vapour cavities leads to deformation and pit formation and, aided by corrosion, heavy metal loss soon occurs.

REMEDIAL MEASURES: One solution would be to change either the volume flow by lowering the inlet duct friction losses or change the impeller design. A marginal improvement could be obtained by choosing stainless steel or aluminium bronze impellers but if the cavitating conditions prevail, these materials would also eventually suffer damage.

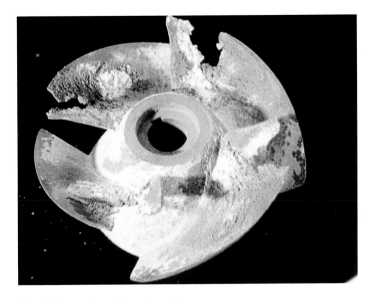

Cavitation of cast iron impeller.

CASE Pu10

PUMP IMPELLER; COPPER-BASE ALLOY; EROSION-CORROSION

INSTALLATION: Seawater pump impeller in a chemical plant cooling system.

DAMAGE: Erosion-corrosion occurred at both the inlet and outlet side of this mixed- flow impeller. Figures 1 and 2 show the attack at the outlet side. In the inlet area the attack was concentrated just after the leading edge and was of a local impingement type.

At the outlet edge the attack was fairly even in depth and started about 30-50mm upstream from the trailing edge. On the back side (pressure side) of the impeller, a similar type of attack can be seen, as at the outlet end of the vanes.

CONDITIONS: Seawater.

MATERIAL: Not given - apparently copper-base alloy.

CAUSE OF FAILURE: The system head (pressure) relative to the pump design has been too low. This results in entry and exit losses at the impeller which cause severe flow disturbances resulting leading to erosion-corrosion, as observed on the impeller.

REMEDIAL MEASURES: Obtain better balance between system head and pump by (1) changing the impeller diameter; (2) throttling the exit valve of the pump (not the inlet valve), and (3) changing the rotational speed.

Figure 1: Erosion-corrosion at the edge of copper-alloy impeller.

Figure 2: General appearance of corrosion damage on copper-alloy impeller.

CASE Pu11

PUMP IMPELLER; Ni-RESIST CAST IRON; EROSION-CORROSION

INSTALLATION: Seawater pump impeller from a 2-stage submersible pump in a power station in West Africa.

DAMAGE: The impeller has suffered general thinning and penetration due to erosion corrosion with some evidence of cavitation (see Fig.).

CONDITIONS: Seawater with sulphides present.

MATERIAL: Impeller - Ni-Resist Type D-2
            Casing - Ni-Resist Type D-2
            Shaft - Stainless steel

REMEDIAL MEASURES: Replaced with materials with higher resistance to flowing seawater, i.e. nickel-aluminium bronze or stainless steel Type 316.

COMMENT: Ni-Resist has good resistance to static and low velocity seawater but suffers erosion-corrosion at high velocities. It is not normally used for pump impellers but is often used for casings, discharge pipes, etc. It can provide useful cathodic protection to stainless steel shafts and impellers when the pump is stationary.

Erosion-corrosion on Ni-Resist impeller.

CASE Pu12

PROPELLER SHAFT; TYPE 316 STAINLESS STEEL; CREVICE CORROSION

INSTALLATION: Ship's propulsion system.

DAMAGE: Crevice corrosion of the propeller shaft (Fig. 1) beneath the inner out-board bearing (Fig. 2).

MATERIAL: Austenitic stainless steel AISI 316 Outboard bearing AISI 316, rubber lined.

CONDITIONS: The propulsion system had been exposed to seawater of ambient temperature with frequent idle periods, sometimes prolonged.

CAUSE OF DAMAGE: Stainless steels of the 304 and 316 series are susceptible to crevice corrosion, initially caused by the formation of a deaeration cell between the inside of the crevice and the surrounding part of the system. The degree of susceptibility depends upon flow conditions and crevice geometry. When the ship is in normal use, sufficient fresh aerated seawater is supplied to the inner side of the bearing to prevent crevice corrosion from occurring. However, when the ship is lying idle, very narrow crevices are formed between the propeller shaft and the inner rubber lining of the bearing.

As well as detrimental crevice geometry, the diffusion of oxygen to the inner side of the crevice will be severely restricted in stagnant conditions. These conditions lead to the initiation and propagation of crevice corrosion.

REMEDIAL MEASURES: Application of cathodic protection by sacrificial anodes in the vicinity of the bearing or the use of a more resistant material.

COMMENTS: This type of attack is typical for conventional stainless steels in seawater when crevices are present. Vertical pumps often have non-metallic seawater-lubricated bearings and can suffer crevice corrosion in the same way as this shaft.

Figure 1: Crevice corrosion of AISI 316 propeller shaft.

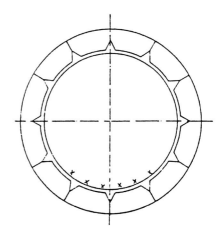

Figure 2: Section through shaft and bearing.

CASE Pu13

SHIP'S PROPELLER; MANGANESE BRONZE; STRESS CORROSION CRACKING (SCC)

INSTALLATION: Ship's propeller.

DAMAGE: Cracks at the blade roots.

MATERIAL: Modified manganese bronze (59Cu - 1Mn - 0.1Fe - 1Al - bal. Zn - other 1.5)

CONDITIONS: The propeller is normally in seawater of ambient temperature and is connected to a carbon steel propeller shaft, which is isolated from seawater.

CAUSE OF DAMAGE: Stress corrosion cracks had occurred (Fig. 1), the cracks propagating mainly from the surface of the hub inwards. The presence of some small porosity was also detected (Fig. 2).

Large internal stresses may be induced by any of the following:

(1) Heating the hub locally to remove the propeller for repair; (2) Pressing the hub too severely on the propeller shaft; (3) Insufficient quality control during fabrication procedure.

If such stresses are induced in the propeller stress corrosion cracking can occur in subsequent use.

REMEDIAL MEASURES: Use of a less SCC-susceptible material— and/or reduction of internal stresses from fabrication, mounting, repair etc.

COMMENT: The most common cause of stress in a propeller hub is non-uniform heating of the hub for removal of the propeller.

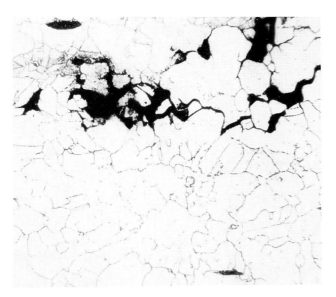

Figure 1: SCC of modified Mn-bronze-propeller.

Figure 2: Detail of Fig. 1.

VALVES

Valves are components in a seawater system which control flow. They are used for shut-off, non-return or throttling purposes. When used in a throttling mode, they are often exposed to severe turbulence both within, and downstream from, the valve.

The most commonly used types of valve in seawater systems are butterfly, globe, ball and gate valves. They consist of three main components, namely the valve body, the stem and the disc (butterfly and globe), ball or gate.

Examples of corrosion commonly experienced in valves are provided in the following cases and cover most of the valve types and components mentioned above.

Stainless steels are often used in valves for components such as valve stems. Depending on the grade chosen and cathodic protection provided, these can suffer pitting and crevice corrosion - several examples of this are described.

Because of the turbulence generated in valves, erosion-corrosion can occur on susceptible alloys such as copper base alloys. Examples of this type of corrosion are given.

Brasses are often used for valve components such as stems and seats and can suffer dezincification unless cathodically protected as, for example, by a ferrous valve body. Dealuminification - a similar phenomenon - can also occur in aluminium bronzes. Examples of selective corrosion are described for both brasses and aluminium bronzes.

Coatings are sometimes used to protect valve parts and two cases of breakdown of metallic coatings - chromium plate and electroless nickel plate - are given.

CASE V1

VALVE STEM; STAINLESS STEEL; CREVICE CORROSION

INSTALLATION: Valve stem from a valve fitted into a non-ferrous system carrying seawater.

DAMAGE: Severe corrosion had occurred where the valve stem is connected to the valve body (see Fig.).

MATERIALS: Valve stem - martensitic stainless steel, AISI 431. Valve - bronze

CONDITIONS: Seawater of ambient temperature.

CAUSE OF DAMAGE: Generally martensitic stainless steel is not resistant to seawater without additional protective measures. In this case the corrosion susceptibility has been greatly increased by the galvanic contact between the valve and the stem, the stem being the active part of the bimetallic combination.

REMEDIAL MEASURES: The stem should have been constructed from a material that is compatible with the rest of the valve and which has adequate strength. Some design changes may be necessary as it is difficult to select more corrosion resistant materials that have the same strength.

COMMENT: Martensitic stainless steels are often chosen for valve parts because of their high strength and low cost (relative to other stainless steels). When used in seawater, failure by pitting and crevice corrosion is probable. Use of this alloy in seawater valve applications should be avoided.

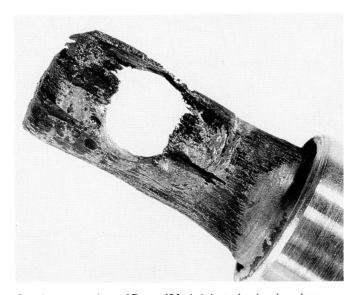

Crevice corrosion of Type 431 stainless steel valve stem.

CASE V2

VALVE STEM; ALPHA BRASS; DEZINCIFICATION

INSTALLATION: The stem of a bronze valve sited in a non-ferrous seawater piping system.

DAMAGE: Severe wastage of valve stem (Figs. 1 and 2).

MATERIALS: Uninhibited-brass shaft; Piping system - CuNi10Fe

CONDITIONS: Flowing seawater of ambient temperature, nominal flow velocity in piping system 1.5 m/sec; prolonged idle periods.

CAUSE OF DAMAGE: Dezincification of the brass occurred. The porous copper remains are very weak and easily damaged by turbulent flowing seawater and wear. As can be seen in Fig. 2 the most severe wastage has occurred on the threaded portion of the shaft.

REMEDIAL MEASURES: Alternative choice of material which should be matched to the other parts of the valve, as well as to the piping system.

COMMENT: Brasses - copper-zinc alloys - are prone to dezincification in seawater. The single phase alpha brasses can be inhibited against dezincification by the addition of a small amount (0.02 - 0.05%) of arsenic. The brass used in this valve was not inhibited and should not have been used in this application.

Figure 1: Wastage of brass valve stem.

Figure 2: Severe wastage at the threaded portion of the stem.

CASE V3

BALL VALVE; CHROMIUM PLATE; GALVANIC CORROSION

INSTALLATION: Ball valve in ship's cargo/ballast system.

DAMAGE: Sticking of the ball valve was observed and after stripping the valve it was found that the stem was corroded and that rust had reduced the clearances thus causing the jamming.

MATERIALS: Cast steel valve body. Rulon (a filled PTFE) seats. AISI 316 stainless steel ball. Chromium plated mild steel stem.

CONDITIONS: The valve had been in service on a vessel for two years and during that time had been subjected to approximately equal lengths of service in lubricating oil products and seawater ballast.

CAUSE OF DAMAGE: Corrosion of the chromium plated mild steel stem adjacent to the stainless steel ball had occurred. Chromium plating is noble to the underlying mild steel and therefore in a seawater environment corrosion of the mild steel will occur at the many defects that are present in chromium plating.

The damage had probably been intensified by the stainless steel ball adjacent to the stem. The corrosion at defects in the chromium plating is shown in the Fig.
Corrosion of the steel adjacent to the stainless steel ball does not occur since the two are insulated by non-conducting Rulon seals.

REMEDIAL MEASURES: The stem should be made from a stainless steel compatible with the ball, i.e. a steel to AISI 316.

Corrosion of chromium plated mild steel stem.

CASE V4

VALVE STEM: STAINLESS STEEL: PITTING

INSTALLATION: Cooling system in a 23,000 dwt container ship carrylng seawater up to 35°C at 3 m/sec.

DAMAGE: The stem of the throttle valve was attacked locally by pitting. Figure 1 shows the distribution of the damage whilst Figs. 2 and 3 are close-ups of the corroded areas.

MATERIALS: Stem - Fe - 17Cr - 2Ni stainless steel. Valve body - tin bronze

CAUSE OF DAMAGE: This type of stainless steel has poor resistance to pitting and crevice corrosion in seawater. An additional factor was the bronze valve body in contact with the shaft - this would provide a large cathodic area to accelerate attack on the stainless steel when pitting initiated.

REMEDIAL MEASURES: Change the material of the shaft to either aluminium bronze (Cu - 10Al - 3Ni - 2Fe) or a higher alloy stainless steel with improved resistance to pitting in seawater (Fe - 25Cr - 5Ni - 1Mo).

Figure 1: Distribution of damage on stainless steel valve stem.

Figure 2: Corroded area of stainless steel valve stem.

Figure 3: Corroded area of stainless steel valve stem.

CASE V5

VALVE BODY: ALPHA-BETA (60/40) BRASS; DEZINCIFICATION

INSTALLATION: Deaeration valve in seawater system.

DAMAGE: Frequent rupturing of valves.

MATERIALS: Valves: brass (CuZn40Pb2 - werkst. nr. 2.0402)
Piping system: CuNi10Fe (werkst. nr. 2.0872).

CONDITIONS: Intermittently, seawater of ambient temperature is extracted from the system through the valves.

CAUSE OF DAMAGE: The brass is of a two-phase type, without addition of inhibiting substances such as arsenic. This type of brass is very susceptible to dezincification in seawater, as the copper-rich α-crystals are more noble than the zinc-rich β-crystals.

The microstructure clearly shows the dezincification process progressing deeply into the cross-section of the material, thereby greatly increasing the susceptibility to rupture by loss-of mechanical strength (Figs. 1 and 2).

REMEDIAL MEASURES: The valves should be replaced using more corrosion-resistant materials such as nickel aluminium bronze or gunmetal.

COMMENT: This type of brass is very prone to dezincify in seawater. It is frequently used for seawater valve components mainly because it is inexpensive and very easy to machine. It was often used for internals in cast iron valves where the cathodic protection provided by the valve body enabled it to provide reasonable service. However, when used on its own without such protection, failure is rapid. The rate of dezincification is reduced if the alloy contains tin (e.g. naval brass 60Cu - 39Zn - 1Sn). However, the improvement obtained is not usually sufficient for valve applications.

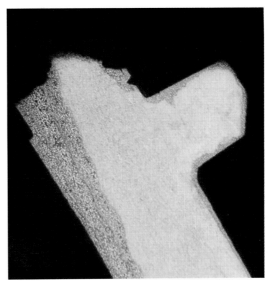

Figure 1: Dezincified area on alpha-beta brass valve body.

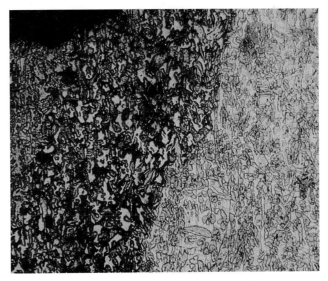

Figure 2: Microstructure of alpha-beta brass valve body showing preferential attack and dezincification of the beta-phase.

CASE V6

VALVE BODY; ALPHA-BETA (60/40) BRASS; DEZINCIFICATION

DAMAGE: Leakage of valves.

MATERIALS: Valve: As specified G-CuSn10Zn (werkst. nr.2. 1086.01).
Valve: As-supplied brass.
Piping system: CuNi30Fe (werkst. nr. 2.0882).

CONDITIONS: The valve operated in seawater of ambient temperature, flowing intermittently at a velocity of 1.5 m/sec over a period of three months.

CAUSE OF DAMAGE: A number of valves showed a red discoloration. The surface was very porous (Fig. 1).

In the shaft of a valve several cracks were found. The microstructure shows two kinds of crystals, one of which has been preferentially attacked (Fig. 2).Chemical analysis proved the material to be brass instead of the specified bronze. The damage is caused by dezincification of the brass; the brass type involved is of the (α + β) type, containing copper-rich α - crystals, next to zinc-rich crystals. The β - crystals are more active and dissolve preferentially.

REMEDIAL MEASURES: Changing the valves to the specified material.

COMMENT: See CASE V5.

Figure 1: Dezincified area on alpha-beta brass valve body.

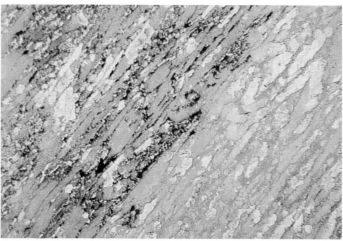

Figure 2: Preferential attack and dezincification of beta-phase.

CASE V7

VALVE BODY; TITANIUM; GENERAL CORROSION

INSTALLATION: Shipboard electrochlorination unit - main condenser system.

DAMAGE: Severe general corrosion of titanium (see Fig.).

CAUSE OF DAMAGE: Caused by failure of insulation on the platinised titanium anode leading to polarisation of the valve components.

REMEDIAL MEASURES: Ensure the anode is fully insulated from the remainder of the system.

REFERENCE: E.B. Shone and G. C. Grim: in 25 Years Experience with Seawater Cooled Heat Transfer Equipment in the Shell Fleet. Transactions of the Institute of Marine Engineers - 1986, 98, Paper 11.

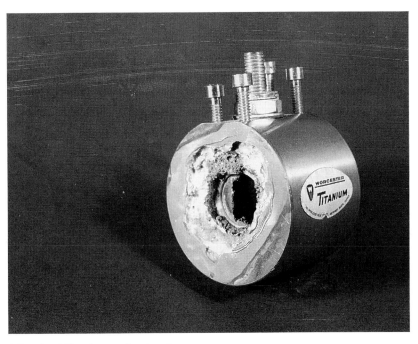

Attack of titanium valve body.

CASE V8

VALVE BODY AND DISC; GUNMETAL AND NAVAL BRASS; EROSION-CORROSION AND DEZINCIFICATION

INSTALLATION: Globe valve - ship's seawater system.

MATERIALS: Valve body, disc seat, stuffing box - gunmetal (85Cu - 5Sn - 5Zn - 5Pb).

Stem & gland - naval brass (62Cu - 1Sn - 37Zn).

CONDITIONS: Turbulent seawater.

CAUSE OF DAMAGE: The seat, disc and part of the body adjacent to the seat had been damaged by severe erosion-corrosion of the type associated with fast flowing turbulent seawater. The damage had probably been intensified because the valve had been throttled in service and this had caused increased turbulence in the valve.

Galvanic corrosion could have accelerated the dezincification of the stem and naval brass is not immune to dezincification in seawater.

REMEDIAL MEASURES: Use materials resistant to turbulent flow, e.g. nickel-copper Alloy 400 (70Ni-40Cu) for the disc, disc-seat and stem and nickel aluminium bronze (Cu-9Al-5Ni-5Fe) for the valve body.

REFERENCE: E.B. Shone: in Problems in Seawater Circulating Systems. British Corrosion Journal 1974, 9, (1), 32-38.

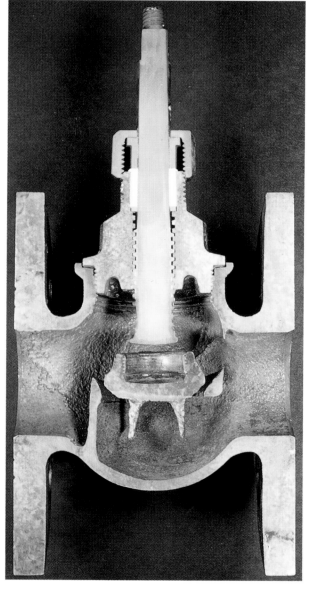

Figure 1: Erosion-corrosion of gunmetal (body and disc) and brass valve stem.

CASE V9

VALVE DISC; TIN BRONZE; EROSION-CORROSION

INSTALLATION: Valve in seawater piping system.

DAMAGE: Leakage of valve.

MATERIAL: Valve: Bronze GCuSn10Zn (werkst.nr. 2.086.01)
Piping system: CuNi10Fe (werkst.nr. 2.0872)

CONDITIONS: Seawater of ambient temperature at a velocity of approximately 1.5 m/sec flowing through the pipes. The system had been installed for about 2 years.

CAUSE OF DAMAGE: The attack was seen in the form of radial grooves, which initiated at the side of the valve (Fig. 1). This kind of attack has originated by erosion-corrosion, which in turn is caused by leakage of water along irregularities in the surface of the valve. The valve material was found to be porous, due to insufficient control of the casting process (Fig. 2).

REMEDIAL MEASURES: Better control of the casting process whereby the presence of pores in the cast material is obviated. A further improvement could be a homogenising heat-treatment, decreasing the susceptibility of micro-galvanic corrosion caused by segregation.

Figure 1: Erosion-corrosion of valve seat.

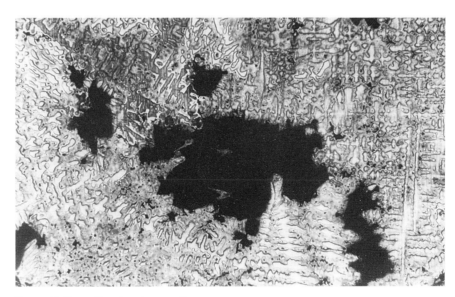

Figure 2: Porosity of valve casting.

CASE V10

BALL VALVE BALL; AUSTENITIC STAINLESS STEEL; CREVICE CORROSION

INSTALLATION: Ball valve ball in a seawater system with stainless steel pipes.

DAMAGE: After 1-2 years of service the ball which had been in intermittent service in clean, filtered seawater, showed crevice corrosion under the O-ring (see Fig.).

MATERIALS: Ball - DIN W. no. 1.4435 (X 2CrNiMo 18/12/ 2.5) stainless steel. Piping - same material.

CONDITIONS: The valve was exposed to clean filtered Mediterranean seawater.

CAUSE OF DAMAGE: This grade of stainless steel (equivalent to UNS S31600) is known to be prone to pitting and, in particular, crevice corrosion in seawater. These types of corrosion can be prevented by application of cathodic protection which lowers the potential from the free corrosion value in seawater.

REMEDIAL ACTION: 1. Cathodic protection was applied by inserting annular rings of mild steel between flanges to act as sacrificial anodes. An alternative way of providing protection would be to connect a length of heavy walled mild steel pipe to the valve. An impressed current cathodic protection system would be possible but is too expensive in most cases.

2. Use seawater resistant materials such as titanium - this is also too expensive in most cases.

COMMENT: Although pitting and crevice corrosion of stainless steels — such as used in this case — are well known, these materials are still often used under conditions where corrosion is likely. They can be used successfully if cathodic protection can be provided. Where this cannot be provided, more resistant alloys should be used.

Crevice-corrosion of Type 316 stainless steel ball valve.

CASE V11

VALVE DISC; NICKEL PLATING ON CAST IRON; PITTING

INSTALLATION: Valve disc in a butterfly valve in a ships seawater system.

DAMAGE: Local pitting in nickel plated cast iron disc (see Fig.).

MATERIALS: Electroless nickel plate (20 microns) on cast iron.

CONDITIONS: Flowing seawater.

CAUSE OF DAMAGE: When nickel plating is applied directly on top of an as-cast surface, porosity and defects are inevitable. The nickel layer is more noble than the substrate metal and this will give rise to rather heavy corrosion attack (galvanic acceleration) at defects as in this case after only 3 months of operation.

REMEDIAL MEASURES: Use substitute materials like rubber-coated cast iron or inherently corrosion resistant material like Ni-Al Bronze.

COMMENT: Thin coatings of materials such as nickel or, as in this case, a nickel-phosphorus alloy, do not provide reliable resistance to corrosion in seawater.

Localized pitting at defects in electroless nickel plate.

CASE V12

VALVE DISC; CAST STEEL; FLOW ACCELERATED GENERAL CORROSION

INSTALLATION: Ship's seawater system - butterfly valve.

DAMAGE: Severe corrosion of the valve disc. (See Fig.)

MATERIALS: Valve body - cast steel.
Valve disc - cast steel
Valve seat (on disc) - stainless steel
weld overlay.

CONDITIONS: Flowing seawater.

CAUSE OF DAMAGE: Corrosion resistance of cast steel in flowing seawater is poor and on this valve disc, corrosion is accelerated by the cathode provided by the stainless steel weld deposit.

REMEDIAL MEASURES: Use materials of construction with good resistance to flowing seawater such as nickel aluminium bronze (Cu-9Al-5Ni-5Fe).

COMMENT: Corrosion of steel is markedly accelerated in flowing compared with static seawater. Use of unprotected carbon steel in flowing seawater requires generous corrosion allowances which, in most valve applications is impracticable.

REFERENCE: E.B. Shone: in Problems in Seawater Circulating Systems. British Corrosion Journal 1974, 9, (1), 32-38.

Figure 1: Corroded valve disc from a butterfly valve showing preferential attack on the cast steel base and protection of the stainless steel seat.

CASE V13

VALVE DISC; ALUMINIUM BRONZE; DEALUMINIFICATION

INSTALLATION: Ships seawater system - butterfly valve - Fig. 1.

DAMAGE: Severe general dealuminification of the valve disc is evident in Fig. 2 (shown on p. 80).

MATERIALS: Valve body - rubber-lined cast iron. Disc - manganese-aluminium bronze (BS1400 CMA1)

CONDITIONS: Flowing seawater.

CAUSE OF DAMAGE: Some aluminium bronzes are prone to dealuminification in seawater.

The use of rubber linings in this valve has effectively removed any cathodic protection that the valve disc would have obtained from the valve body and which would have greatly reduced the attack.

REMEDIAL MEASURES: Use an aluminium bronze with greater resistance to dealuminification - nickel aluminium bronze (BS 1400 AB2 (Cu-9Al-5Ni-5Fe)).

REFERENCE: E.B. Shone: in Problems in Seawater Circulating Systems. British Corrosion Journal 1974, 9, (1), 32-38.

Figure 1: General appearance of butterfly valve showing corrosion of the disc.

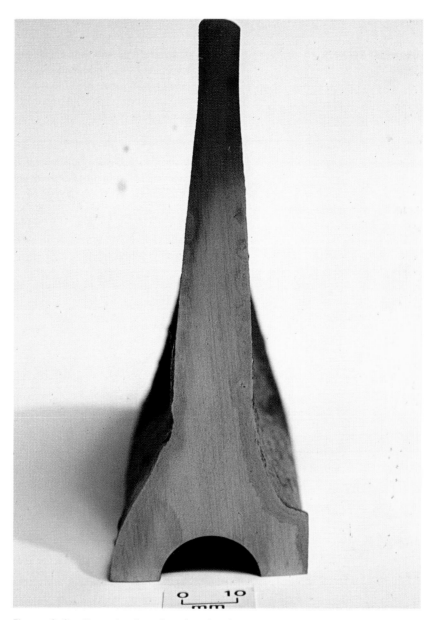

Figure 2: Section of valve disc showing layer of dealuminification.

CASE V14

VALVE DISC; ALUMINIUM BRONZE; DEALUMINIFICATION

INSTALLATION: Auxiliary valve in power station cooling water system.

DAMAGE: Corrosion of the bronze, taking the form of dealuminification, had occurred in the valve disc (see Fig.).

CONDITIONS: Brackish cooling water.

CAUSE OF DAMAGE: Some types of aluminium bronze are prone to dealuminification in seawater and brackish waters. More resistant grades should be chosen for these environments, e.g. BS1400 AB2 (Cu-9Al-5Ni-5Fe).

REMEDIAL MEASURES: In the case of such small valves, all-coated or non-metallic valves can be used.

Corrosion of a valve disc due to dealuminification.

CASE V15

VALVE DISC; CAST IRONS; GENERAL CORROSION ACCELERATED BY FLOW

INSTALLATION: Main condenser discharge valves in power station cooling water system.

DAMAGE: Rapid metal loss on the upstream sealing edges of the butterfly valve disc. Figure 1 shows grey cast iron valve disc. Figure 2 shows Ni-Resist iron valve disc.

MATERIAL: Cast iron.

CONDITIONS: Brackish cooling water with high suspended solids content.

CAUSE OF DAMAGE: These examples highlight the important influence of cooling water velocity and thus the need to minimise flow disturbances.

REMEDIAL ACTION: In situations where turbulence cannot be avoided, consideration to the use of other materials, e.g. non-metallics, should be given.

COMMENT: See comment in CASE V12 - cast iron is affected by flow rate in a similar way to steel.

Figure 2: Erosion-corrosion of Ni-Resist valve disc.

Figure 1: Erosion-corrosion of cast iron valve disc.

CASE V16

VALVE DISC; ALUMINIUM BRONZE; EROSION-CORROSION

INSTALLATION: Butterfly valve disc from a power station seawater cooling system.

DAMAGE: After approximately 5 years service in a rubber-lined S.G. iron valve body, the disc suffered severe corrosion. The corrosive attack was most severe around the edges of the disc and in areas of high water velocity and was typical of erosion-corrosion - Fig. 1.

CAUSE OF DAMAGE: The disc was made from a copper aluminium alloy containing magnesium. It did not conform to any known standard. Metallographic examination of areas close to those suffering erosion-corrosion showed selective phase attack - Fig. 2. This would make the alloy more susceptible to erosion-corrosion attack.

Presence of sulphides in the surface film was confirmed by the iodine/sodium azide test. This would also make the alloy more susceptible to erosion-corrosion as sulphides have a marked effect on corrosion of aluminium bronzes in seawater.

Originally the valve had been connected to steel pipes which would provide some cathodic protection. This protection was removed when the steel was replaced by plastics pipes.

REMEDIAL MEASURES: Body - rubber-lined S.G. iron (3-4mm rubber lining); Disc - nickel aluminium bronze (BS1400 AB2C) or rubber lined disc; Spindle or fasteners (exposed to seawater) - Type 316 stainless steel (min grade) or nickel-copper Alloys 400 or 500; Spindle (sealed from seawater contact)- stainless steel Type 431.

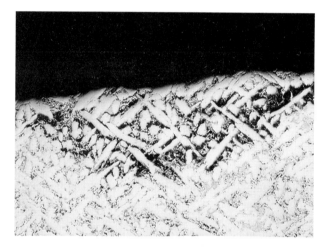

Figure 2: Dealuminification of aluminium bronze.

Figure 1: Erosion-corrosion of aluminium bronze valve disc.

HEAT EXCHANGERS

The majority of seawater cooled heat exchangers are of the tube and shell type with the seawater flowing inside the tubes. In this section on corrosion in seawater cooled heat exchangers a number of case histories are presented following the flow of the cooling water into the water box, passing the tube plate into the tube inlets and through the tubes. Some corrosion problems in plate heat exchangers have also been included.

A number of problems on the shell (process) side have also been described. Although not strictly within the scope of this publication - marine corrosion - these phenomena have nevertheless been included as they are of great practical interest for many marine applications.

The extent and type of corrosion problems encountered are of course dependent on the materials selected and the specific operating conditions for each installation. A short description of the most frequent corrosion problems in components of tube and shell heat exchangers is given below.

Water boxes: Most corrosion problems in water boxes are due to localized attack, usually galvanically accelerated by the metallic contact with the more noble tube plate and tube materials. A defective coating is often a contributing factor.

Tube plates: As in the case of water boxes most problems in tube plates are due to galvanically accelerated localized attack, usually erosion-corrosion, crevice corrosion or dealloying.

Tubes: With copper alloys one of the most common problems is erosion-corrosion at the tube inlet. This may also arise along the tube length if foreign objects are lodged in the tubes. Flow velocity - or rather turbulence - as well as cooling water quality are decisive factors in this respect. A high content of suspended solids will increase the risks for erosion-corrosion problems.

On the other hand a low flow velocity will increase the risk of localized attack under deposits, especially at high temperatures on the shell side. Formation of cathodic films on the cooling water side has caused pitting along scratches or other defects in the film. Stagnant conditions may also cause stress corrosion cracking in highly stressed areas of some copper alloys under certain conditions.

For conventional stainless steel grades nearly all corrosion problems are due to low flow conditions, e.g. under deposits or in crevices resulting in localized attack. At high seawater temperatures these grades are prone not only to pitting and crevice corrosion but also to stress corrosion cracking.

CASE HE 1

WATER BOX; CAST IRON; COATING FAILURE

INSTALLATION: Turbo generator condenser on a ship.

DAMAGE: Corrosion of the waterbox had occurred and the urethane coating was discoloured by rust around the bolts and in corners (see Fig.).

MATERIAL: The waterbox was made from grey cast iron.

CONDITIONS: The unit was cooled by seawater.

CAUSE OF DAMAGE: The corrosion of the waterbox was caused by seawater gaining access to its surface at defects in the coating. The defects in the coating may have arisen due to poor preparation of the metal prior to coating and/or poor application of the coating.

REMEDIAL MEASURES: The defective areas were cleaned and following adequate surface preparation coating repairs were undertaken.

COMMENTS: Failure of protective coatings do occur quite frequently but usually only small areas are affected and local repair is sufficient.

The main cause of serious coating failure is usually poor surface preparation of the metal prior to coating

In the case of waterboxes it is important to carry out coating inspection regularly as any breakdown can result in rapid metal loss due to the presence of more noble metals (tube plates and tubes) in the system.

Figure 1: Local breakdown of waterbox coating.

CASE HE 2

TUBE PLATE; NAVAL BRASS; EROSION-CORROSION

INSTALLATION: Heat exchanger on a ship.

DAMAGE: The damage took the form of localised pitting of the tube plate, as shown (see Fig.).

MATERIAL: The tube plate was made to BS 2875-CZ112 (naval brass) and the tubes to BS 2871-CN102 (90:10 copper-nickel).

CONDITIONS: The unit was cooled by seawater flowing at a nominal velocity of 1.5 m/s.

CAUSE OF DAMAGE: Fast flowing, turbulent seawater had caused pitting damage on the tube plate by erosion-corrosion. The adverse flow conditions that prevailed in the waterbox must be associated with design inadequacies.

REMEDIAL MEASURES:Cathodic protection of the tube plates on similar units was applied by fitting soft iron sacrificial anodes. Care was taken to ensure that the anodes did not significantly interfere with the flow pattern within the waterbox.

COMMENTS: Tube plate damage, such as this, does occur quite frequently. There are several methods available for overcoming this problem in addition to the actions taken in this case such as:
(a) improving the design of the waterbox inlet to prevent excessive turbulence;
(b) adding ferrous sulphate to the incoming cooling seawater; the amount of iron salt addition should be similar to that used to inhibit erosion-corrosion on aluminium brass;
(c) coating the tube plate with a suitable material, e.g. epoxy compound.

Section of tube plate showing surface damage.

CASE HE 3

TUBE PLATE; NAVAL BRASS; EROSION—CORROSION

INSTALLATION: 120 MW estuarine water cooled steam condenser in a power station.

DAMAGE: Severe metal wastage of both tube plate and tube ends, as shown in Figs. 1a and 1b.

MATERIAL: The tube plate was made to BS 2875-CZ112 (naval brass) and the tubes to BS 2871-CN108 (70:30 copper-nickel).

CONDITIONS: The unit was cooled by estuarine water flowing at a nominal velocity of 2 m/s.

CAUSE OF DAMAGE: Severe erosion-corrosion had occurred primarily due to the very high level of suspended solids in the cooling water.

REMEDIAL MEASURES: The problem was overcome by coating the tube plate with neoprene rubber and fitting nylon inserts into the tube inlets. Later the unit was retubed with a more erosion-corrosion resistant copper-nickel alloy.

COMMENTS: Severe damage such as this is not very common but in the case cited the water did contain an exceptionally high proportion of suspended solids.

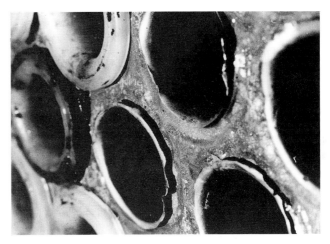

(a)

Figure 1: Section of tube plate showing damage to both tube plate surface and tube ends.
(a) and (b): different views.

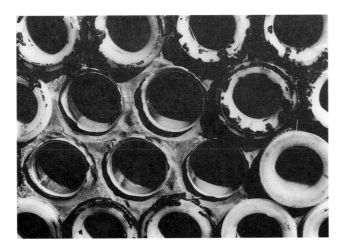

(b)

CASE HE 4

TUBE PLATE; NICKEL ALUMINIUM BRONZE; CREVICE CORROSION

INSTALLATION: Power station gland steam condenser.

DAMAGE: Metal wastage of the tube plate in the vicinity of the waterbox - tube plate joint, as shown in the Fig.

MATERIAL: The tube plate was made to BS 2875-CA105 (nickel aluminium bronze) and the tubes to ASTM B 338 grade 2 (titanium).

CONDITIONS: The unit was cooled by estuarine water flowing at a nominal velocity of 2 m/s.

CAUSE OF DAMAGE: Stagnant water present between the water box - tube plate joint faces had caused crevice corrosion of the tube plate flange face. The resulting corrosion products that formed on the tube plate surface adjacent to the joint produced crevice conditions on the tube plate surface at the periphery of the tube bundle and led to an even greater metal loss.

REMEDIAL MEASURES: The extent of the waterbox-tube plate crevice was reduced by fitting a full face gasket between the metal surfaces.

Portion of tube plate showing crevice corrosion damage.

CASE HE 5

TUBE PLATE; STAINLESS STEEL; CREVICE CORROSION

INSTALLATION: 600 MW brackish water cooled steam condenser in a power station.

DAMAGE: Crevice corrosion of tube plate, as shown in Figs. 1 and 2.

MATERIAL: The tube plate was made of carbon steel with a surface cladding of 4 mm thick AISI type 316 ("stainless" steel). The tubes were titanium.

CONDITIONS: The unit was cooled by brackish water.

CAUSE OF DAMAGE: Cooling water had seeped into the crevice between the tube and tube plate causing corrosion of the carbon steel. This had been intensified by the presence of more noble metals.
The corrosion products had been deposited on the surface of the "stainless" steel (Fig. 1) causing crevice corrosion (Fig. 2).

REMEDIAL MEASURES: The problem was overcome by coating the tube plate with epoxy compound.

COMMENTS: This combination of materials is not very common in heat exchangers. The problem may have been avoided by the use of a thicker tube plate cladding, preferably with a higher molybdenum content.

Titanium cladding, seal welded to the tube ends, would be even better.

Figure 1: Build up of corrosion deposits arising from attack of carbon steel substrate.

Figure 2: As above after cleaning showing attack of stainless steel clad in areas previously covered by deposits.

CASE HE 6

TUBE PLATE; ALUMINIUM BRONZE; CREVICE CORROSION

INSTALLATION: Evaporator condenser on a ship.

DAMAGE: Severe corrosion of the tube plate (Figs. 1 and 2) adjacent to all joint areas together with some general corrosion on the tube plate surface.

MATERIAL: The tube plate was made to BS 2875-CA106 (aluminium bronze) and the tubes were titanium.

CONDITIONS: The condenser was cooled by inshore and deep seawater. It was intermittently operated allowing seawater to remain stagnant within the tubes for up to about one week.

CAUSE OF DAMAGE: Main cause of damage believed to be due to galvanically accelerated crevice corrosion occurring during periods of seawater stagnancy. The attack (up to 5 mm in 9 months service) occurred outside the actual crevice area.

Up to 0.5 mm was lost from general tube plate surface with some preferential attack also evident adjacent to tubes.

REMEDIAL MEASURES: None taken as the vessel left service.

Figure 1: General view of tube plate.

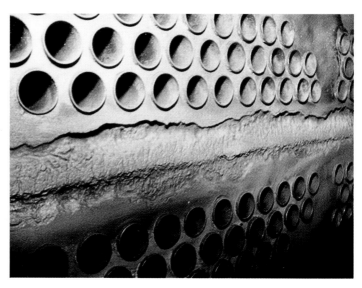

Figure 2: Tube plate damage.

CASE HE 7

TUBES; COPPER-NICKEL; DISSIMILAR METAL CORROSION

INSTALLATION: Lubricating oil cooler used on an offshore platform.

DAMAGE: Perforation of several tubes.

MATERIAL: The tube plate was made to BS 2875-CN102 (90:10 copper-nickel) and the tubes to BS 2871-CN102 (90:10 copper-nickel). A brazing alloy approximating to the American Standard type B Ag2 was used and the water box was BS 1400-5CBl (70:30 copper-zinc brass).

CONDITIONS: Seawater typically at 9° C flowed at a nominal speed of 2.5 m/s through the tubes, cooling oil at an inlet temperature of 69 C.
The cooler had a two pass water flow and one pass oil flow and had been in service for a few months before failure.

CAUSE OF DAMAGE: The tubes were sealed into the tube plates by braze metal which completely covered both faces of the tube plate and had flowed into the tube bores typically to a distance of 5-10 mm (see Fig.). The ends of the tubes protruded from the tube plate by about 1 mm and those ends in the inlet passes were eroded.

Severe pitting and perforation of the copper-nickel had occurred along the tubes at the interface of braze metal and the tubes. The braze metal itself was not corroded.

The potentials of the braze metal and the tube alloy were measured in salt water and potential differences of up to 120 mV were found with the braze metal being cathodic.

This potential difference is considered to be the prime cause of the pitting.

REMEDIAL MEASURES: The tubes affected were removed from service by plugging.

COMMENTS: This failure may have been avoided by using a less noble brazing alloy during construction.

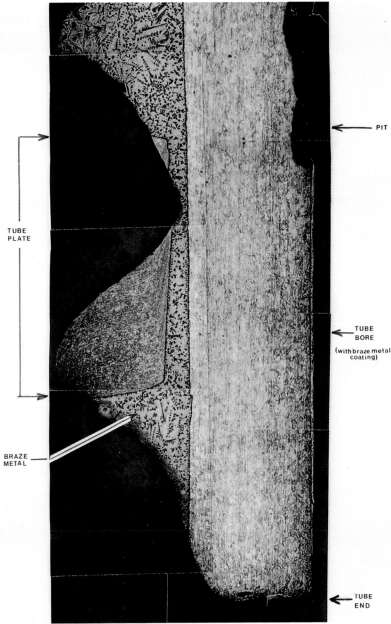

Section through tube end showing metal loss at the edge of the braze metal coating (upper right hand side of picture).

CASE HE 8

TUBES; COPPER-NICKEL; EROSION-CORROSION

INSTALLATION: 120 MW estuarine water cooled steam condenser in a power station.

DAMAGE: Pitting and in some instances, perforation of tube inlets.

MATERIAL: The tubes were made to BS 2871-CN108 (70:30 copper-nickel).

CONDITIONS: The tubes were cooled by estuarine water flowing at a nominal velocity of 2 m/s.

CAUSE OF DAMAGE: Severe erosion-corrosion of the tube inlets had occurred although a normally resistant alloy had been used. The damage, an example of which is shown in the Fig., occurred after 16,000 hours of operation. The reason for this unusually high rate of attack was the very high level of suspended solids in the cooling water (typically 3000 ppm).

REMEDIAL MEASURES: The condenser was retubed with an even more erosion-corrosion resistant copper-nickel alloy containing 64Cu- 30Ni-3Fe-3Mn.

COMMENTS: Erosion-corrosion of heat exchanger tube inlets is quite common. Normally remedial action such as fitting inserts, ferrous sulphate dosing, cathodic protection, coating, etc. is attempted with retubing being the last resort. However, in this case retubing was necessary due to the severity of attack. The limited remaining life of the plant was such that, in this instance, titanium was not used.

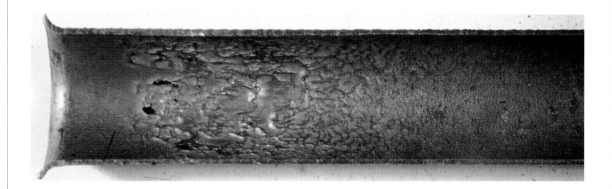

Section through tube at inlet end showing metal loss.

CASE HE 9

TUBES; ALUMINIUM BRASS; EROSION—CORROSION

INSTALLATION: 500 MW seawater cooled steam condenser in a power station.

DAMAGE: Pitting attack of the type associated with erosion-corrosion has occurred at the inlet ends of the tubes (see Fig.).

MATERIAL: The tubes were made to BS 378-CZ110 (aluminium brass).

CONDITIONS: The tubes were cooled by seawater flowing at a nominal velocity of 2 m/s.

CAUSE OF DAMAGE: Fast flowing turbulent seawater had caused the damage to the tube inlets.

REMEDIAL MEASURES: Nylon inserts (150 mm length) were fitted into the tube inlet ends.

COMMENTS: Erosion-corrosion of heat exchanger tube inlets is quite common. Ferrous sulphate dosing or cathodic protection were other possible solutions. If none of these options was successful then retubing with a more resistant alloy, e.g. copper-nickel, would be necessary.

Inlet end of tubes showing erosion-corrosion.

CASE HE 10

TUBES; ALUMINIUM BRASS; EROSION-CORROSION

INSTALLATION: Heat exchanger in a power generating unit.

DAMAGE: Pitting attack of the type associated with erosion-corrosion had occurred at the tube ends (inlet and outlet). The damage was confined to 10-15 cm into the tubes.

MATERIALS: The tubes were made from aluminium brass and the tube plate to AISI 316L (stainless steel). A rubber lining had been applied to the water box.

CONDITIONS: The tubes were cooled by clean seawater flowing at a nominal velocity of 1.8 m/s. A sponge rubber ball tube cleaning system, together with ferrous sulphate dosing, had been used on the unit.

CAUSE OF DAMAGE: The damage, which was essentially erosion-corrosion (see Fig.), had occurred after 3-6 months in service. It had been stimulated by the galvanic couple that existed between the more noble stainless steel and the aluminium brass. Without this stimulation erosion-corrosion would have been unlikely to occur under these operating conditions.

REMEDIAL MEASURES: Sacrificial soft iron anodes were fitted to the tube plate and this was found to be adequate to prevent further damage.

COMMENTS: Failures of this type are unusual in heat exchangers as this combination of materials is uncommon.

Tube ends showing internal pitting.

CASE HE 11

TUBES; ALUMINIUM BRASS; EROSION-CORROSION

INSTALLATION: Ship's condenser with seawater cooling.

DAMAGE: Pitting attack of the type associated with erosion-corrosion had occurred at the inlet end of the tubes (Figs. 1 and 2). The tubes affected were near to the seawater inlet pipe (Figs. 3 and 4, shown on p. 96). In some instances perforation of the tubes had occurred.

MATERIALS: The tubes were made to Werkst. nr. 2.0460 (aluminium brass), the tube plate to DIN CuZn 39 Sn, Werkst. nr. 2.0530 (naval brass) with the water box to CuSn10Zn, Werkst. nr. 2.1087 (bronze) and a plain carbon steel shell.

CONDITIONS: The condenser was cooled by seawater at an ambient temperature of 8-18°C flowing at a nominal velocity of 1.5 m/s. Cathodic protection, in the form of soft iron anodes attached to the tube plate, was applied.

CAUSE OF DAMAGE: Impingement attack (erosion-corrosion) due to high turbulence levels had caused perforation of tubes after about 20,000 hours operation. In the areas which showed attack the flow velocity had been above the nominal design velocity because of a design error —— insufficient attention having been paid to the shape of the water box and the positioning of the seawater inlet pipe.

REMEDIAL MEASURES: The unit was retubed with tubes to BS 2871-CN108 (70:30 copper-nickel). In this instance there was insufficient room to enable either a ferrous sulphate injection system to be installed or additional iron anodes to be fitted into the waterbox. Impingement attack (erosion-corrosion) due to high turbulence levels at the inlet of tubes and the inability of the iron anodes to provide adequate protection is a frequent problem in modern seawater cooled heat exchangers.

Figure 1: Section through inlet end of tube showing pitting and perforation.

Figure 2: Surface profile of damage.

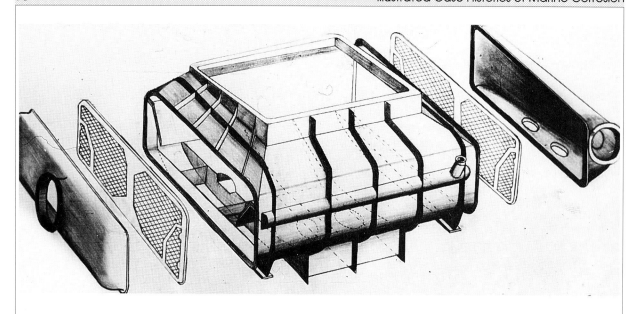

Figure 3: Exploded view of heat exchanger.

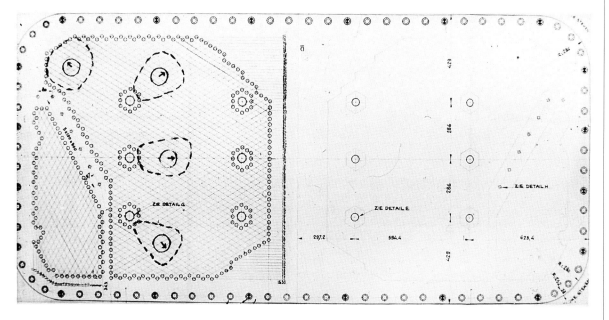

Figure 4: Tube plate diagram.

CASE HE 12

TUBES; ALUMINIUM BRASS; EROSION-CORROSION

INSTALLATION: Seawater cooled condenser.

DAMAGE: Pitting attack of the type associated with erosion-corrosion near the inlet ends of a few tubes.

MATERIALS: The tubes were made from aluminium brass with nylon inlet end inserts fitted.

CONDITIONS: The condenser was cooled by seawater flowing at a nominal velocity of 1.8 m/s.

CAUSE OF DAMAGE: The damage, which took the form of erosion-corrosion, was located immediately downstream of the inlet end inserts on a few tubes (Fig. 1). On inspection it was found that the feathered edge of these inserts had been damaged. The type of damage, reproduced in the laboratory, is shown in Fig. 2.

The distorted inserts had caused turbulent cooling water flow locally resulting in the damage observed.

REMEDIAL MEASURES: The replacement of the damaged inserts overcame the problem.

COMMENTS: This, together with similar experience at other locations, highlights the need for great care during the fitting of inserts into tubes.

Figure 2: Tube cut to reveal damaged edge of nylon insert.

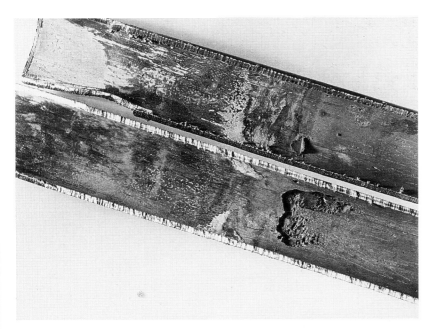

Figure 1: Section through tube near inlet end showing metal loss .

CASE HE 13

TUBES; COPPER-NICKEL; EROSION-CORROSION

INSTALLATION: Estuarine water cooled steam condenser in a power station.

DAMAGE: Very severe localised pitting and perforation of a tube near its inlet end.

MATERIALS: The tube was made to BS 2871-CN108 (66Cu- 30Ni2Fe2Mn) with neoprene coating at the inlet end.

CONDITIONS: The tube was cooled by estuarine water flowing at a nominal velocity of 2 m/s.

CAUSE OF DAMAGE: The condenser tube plate had been coated with neoprene rubber which had been extended into the tube inlets for a short distance (c. 150 mm).

In the failed tube the downstream edge of this coating had become detached. This had caused a disturbance to cooling water flow resulting in localised, but very severe, metal loss.

The Fig. shows the damage viewed from the downstream edge.

The situation was made worse by the high concentration of suspended solids in the cooling water.

REMEDIAL MEASURES: The leaking tube was plugged and an inspection of the other tubes was made.

COMMENTS: This incident demonstrates the problems that can arise when flow disturbances are present in heat exchanger tubes.

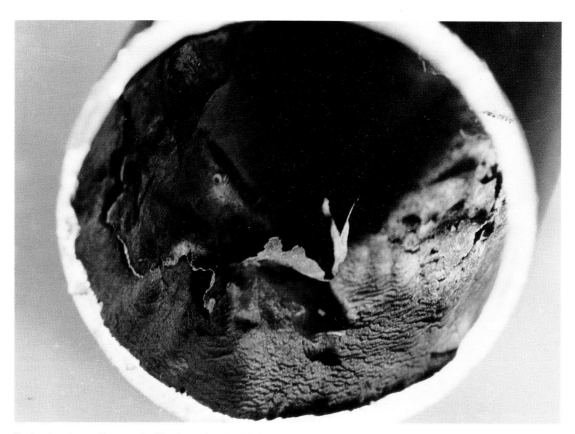

Detached coating and pitted-perforated tube bore.

CASE HE 14

TUBES; ALUMINIUM BRASS; EROSION-CORROSION

INSTALLATION: 500 MW seawater cooled steam condenser in a power station.

DAMAGE: Perforation and pitting of tubes occurred adjacent to obstructions lodged within them.

Figure 1 shows the debris inside a tube and in Fig. 2 (p.100), which is a section through a tube, impingement attack associated with the debris can be clearly seen.

MATERIALS: The tubes were made to BS 378-CZ110 (aluminium brass).

CONDITIONS: The tubes were cooled by seawater flowing at a nominal cooling water velocity of 2 m/s.

CAUSE OF DAMAGE: Localised impingement attack (erosion-corrosion) had been caused by the turbulent water conditions that existed near the obstructions lodged within the tubes (Figs. 1 and 2). The debris responsible was mostly flakes of "rust" which had become detached from the unprotected steel cooling water intake pipework.

REMEDIAL MEASURES: The problem was overcome in two ways: (a) application of a coating to the internal surfaces of the water intake pipework to prevent further corrosion; (b) fitting "reduced throat" plastic inserts into all tube inlets (Fig. 3, shown on p. 100). In this way large pieces of debris were filtered from the system.

COMMENTS: The problem of localised corrosion due to the presence of flow disturbances is quite common in seawater cooling systems.

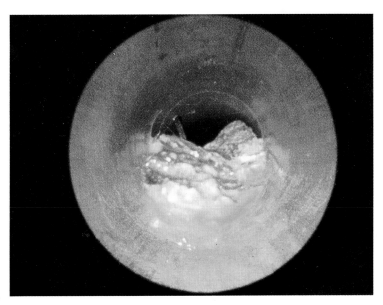

Figure 1: View down tube bore showing debris.

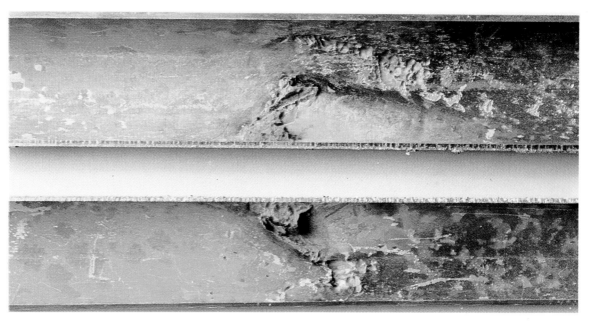

Figure 2: Section through tube showing debris on bore surface and associated erosion-corrosion damage.

Figure 3: View showing part of inlet end tube plate with inserts fitted —note some debris has been "filtered".

CASE HE 15

TUBES; ALUMINIUM BRASS; EROSION-CORROSION

INSTALLATION: Baltic seawater cooled steam condenser in a 700 MW power station. Perforation of a tube adjacent to an obstruction.

MATERIAL: The tube was made from aluminium brass.

CONDITIONS: The condenser was cooled by seawater flowing at a nominal velocity of 1.8 m/s.

CAUSE OF DAMAGE: The condenser had been cleaned with small brushes driven through the tubes by water/air jets. In this case a brush was left in the tube which, on return to service, caused localised cooling water turbulence. The tube became perforated in less than one year. The Fig. shows a section through the tube and the "imprint" of the brush bristles can be seen.

REMEDIAL MEASURES: The leaking tube was plugged and an inspection of other tubes was carried out.

COMMENTS: This incident demonstrates the problems that can arise when flow disturbances are present in heat exchanger tubes.

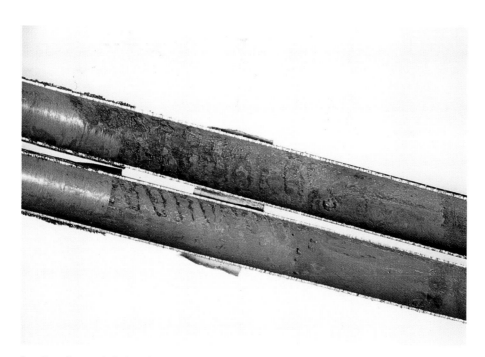

Section through tube showing "imprint" where brush was lodged and a small perforation near by.

CASE HE 16

TUBES; CARBON STEEL; PITTING

INSTALLATION: Oil heater tubes on an offshore platform.

DAMAGE: Pitting and perforation.

MATERIAL: The tubes were made from plain carbon steel.

CONDITIONS: The tubes were found to be leaking before they entered service.

CAUSE OF DAMAGE: It was found that all the tubes were partially covered, inside and outside, by adherent millscale but where this was absent severe rusting and pitting had occured (see Fig.). The intensity of the pitting and corrosion on both the outside and inside of the tubes varied along the tube length.

Corrosion tests and corrosion potential measurements were made which showed the millscale to be highly cathodic relative to the clean steel. This would encourage pitting at discontinuities in the millscale. The corrosion product present in the pits was rich in chloride indicating that brackish water had been present in the tubes.

The source of the brackish water is not known, however, it is probable that it had either entered the tubes during storage or previous hydraulic testing.

REMEDIAL MEASURES: In this case the complete unit was replaced.

COMMENTS: The presence of millscale can, under some circumstances, stimulate pitting corrosion and where it is considered that this may arise consideration should be given to pickling or the use of an inhibitor.

Pitted surface of tube bore (top picture) and external surface (bottom picture) including perforation.

CASE HE 17

TUBES; COPPER-NICKEL; PITTING

INSTALLATION: Ship's main condenser cooled by seawater.

DAMAGE: Pitting and perforation of many tubes (see Fig.).

MATERIALS: The tubes were made to BS 2871-CN108 (66Cu30Ni2Fe2Mn) and the tube plate to BS 2875-CN102 (90:10 copper-nickel).

CONDITIONS: The seawater inlet temperature was about 10-25°C and its nominal velocity was 2 m/s. In this unit the steam temperature was about 40°C.

CAUSE OF DAMAGE: Pitting leading to perforation of 69 tubes (about 1%) was caused by the presence of highly cathodic, by 200 mV, carbon and manganous oxide films on the waterside of the tubes. These films, which were formed during the manufacturing process, promote enhanced localised attack of the alloy. Failures occurred after 3 weeks at sea following an extended precommissioning period.

REMEDIAL MEASURES: The perforated tubes were plugged and the remainder were abrasively cleaned using the carborundum sponge ball method.

COMMENTS: Tubes coated with harmful cathodic films should not be used in seawater service. Such tubes should be cleaned using either the technique mentioned above or a suitable alternative, e.g. acid cleaning, grit blasting.

When clean the tube surface should have a steady state corrosion potential in salt water of not more than 70 mV cathodic relative to freshly abraded tube alloy.

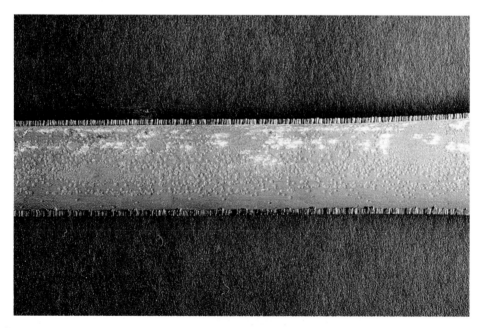

Section through tube showing pitting on bore surface.

CASE HE 18

TUBES; ALUMINIUM BRASS; PITTING

INSTALLATION: 500 MW seawater cooled steam condenser in a power station.

DAMAGE: Pitting of tubes occurred with the pits forming a longitudinal band (Fig. 1). Some had completely penetrated the tube wall (Fig. 2).

MATERIALS: The tubes were made to BS 378-CZ110 (aluminium brass).

CONDITIONS: The condenser was cooled by seawater flowing at a nominal velocity of 2 m/s.

CAUSE OF DAMAGE: The damage was caused by the presence of a highly cathodic manganous oxide film over most of the tube surface on the waterside. The source of the manganese, which is present on occasions in the cooling water, has not been identified. The pitting attack occurred in the lower part of the tube which was shielded from the manganese pollution by a silt layer. This manganese free band was found to be anodic to the rest of the tube surfaces.

REMEDIAL MEASURES: The tubes were cleaned with ultra high pressure air/water jetting.

COMMENTS: In cases such as this the source of the water contamination should be identified and if possible, eliminated. Ferrous sulphate injection may reduce the problem of manganese films.

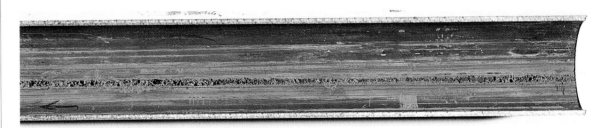

Figure 1: Section through tube showing linear defect on cooling water side.

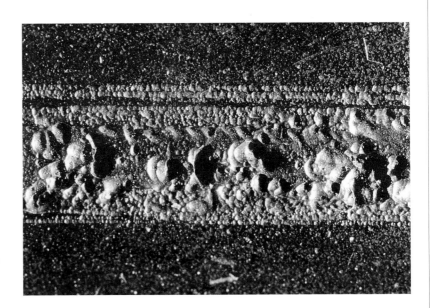

Figure 2: As above at higher magnification showing perforation of tube wall.

CASE HE 19

TUBES; COPPER-NICKEL; PITTING

INSTALLATION: Lubricating oil cooler on a 250,000 dwt oil tanker.

DAMAGE: Several tubes perforated near their outlet ends, the damage occurring in the tops of the tubes.

MATERIALS: The tubes were made to BS 2817-CN108 (66Cu30Ni2Fe2Mn).

CONDITIONS: Lubricating oil at a maximum temperature of 65°C was being cooled by seawater with an inlet temperature of between 10 and 25°C. The designed flow rate of the seawater was 2.5 m/s but at times the flow rate was very low, 0.1 m/s, particularly if partial blockage occurred in the tubes.

CAUSE OF DAMAGE: At low flow rates pockets of air collected in the tops of the tubes shielding them from the cooling water. Under these conditions localised high temperatures occurred in the moist chloride-containing air pockets and rapid corrosion took place. The lower part of the tube which was constantly wetted by cooling water did not corrode.

Typical damage associated with this form of corrosion is shown in the Fig.
This damage was most severe at the outlet end where temperatures were highest and air pockets are most likely to collect. Tube perforation occurred after 9 months in service.

REMEDIAL MEASURES: The failed tubes were plugged and the unit cleaned regularly to avoid debris build up. Low flow cooling water conditions were kept to a minimum.

COMMENTS: This incident has highlighted the problem of low cooling water flow in copper-nickel alloys. Wherever possible cooling water flow rates should not be allowed to fall below about 0.25 m/s.

Section through tube showing pitting of the surface.

CASE HE 20

TUBES; COPPER-NICKEL; PITTING

INSTALLATION: 660 MW seawater cooled steam condenser in a power station.

DAMAGE: Rapid localised pitting corrosion and in some cases perforation generally on the upper parts of the tubes located in the top sections and outer edges of the condenser.

MATERIALS: The tubes were made to BS 2871-CN108 (66Cu-30Ni-2Fe-2Mn).

CONDITIONS: The condenser was cooled by seawater flowing at a nominal velocity of 2 m/s.

CAUSE OF DAMAGE: On occasions severe blockage of tube inlets occurred due to sea weed ingress to the inlet waterbox. The tubes located in the upper parts and outer edges of the condenser, i.e. those most directly affected by steam flow, increased in temperature and some developed leaks, as shown in Fig. 1.

Sections taken through affected tubes showed very patchy surface films on the cooling water side and often copper deposition was observed (Fig. 2). When the deposit was removed many deep pits were revealed (Fig. 3) on the part of the tube facing the steam flow (usually the upper half). The combination of low cooling water flow and overheating has allowed air pockets to form within the tubes leading to the circumstances outlined in the previous case (CASE HE 19).

REMEDIAL MEASURES: All perforated tubes were plugged and steps were taken to prevent sea weed ingress by improved screening so that designed cooling water flow could be maintained in all tubes.

COMMENTS: Copper-nickel alloys of the 70:30 type are particularly prone to these effects and so if conditions cannot be improved an alternative tube material, e.g. titanium, may be required.

Figure 1: Outer edge tubes in condenser (steamside) showing leaks (spots) near the end of one of the tubes.

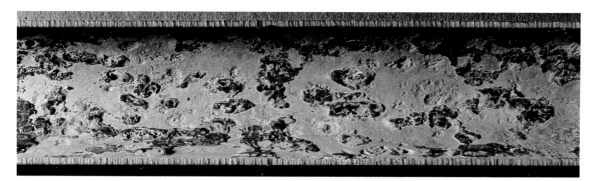

Figure 2: Sectioned tube showing patchy film and copper deposition on tube cooling water surface.

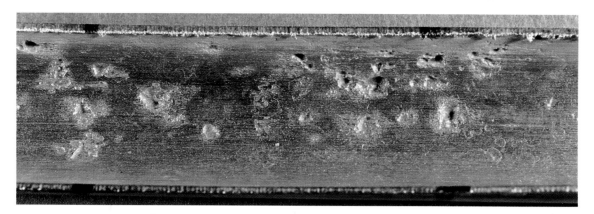

Figure 3: Similar tube to above after cleaning.

CASE HE 21

TUBES; COPPER-NICKEL; PITTING

INSTALLATION: Seawater cooled dump steam condenser in a power station.

DAMAGE: Very rapid localised pitting corrosion on the cooling water side of outer edge tubes near the tube support plates.

MATERIAL: The tubes were made to BS 2871-CN108 (66Cu-30Ni-2Fe-2Mn).

CONDITIONS: The condenser was cooled by sea water flowing at a nominal velocity of 2 m/s.

CAUSE OF DAMAGE: This, relatively small, condenser was used during certain commissioning trials at the power station before the main condensers were available. Often the incoming steam was very hot (up to 110°C). It is suspected that on occasions the cooling water flow rate was less than design. This combination of circumstances led to rapid perforation, i.e. after only about 500 hours, on some of the outermost tubes from the cooling water side by hot spot corrosion (Fig. 1). This type of damage is characterized by very localised deep pits often containing plugs of copper (due to denickelification) such as shown in Fig. 2.

In this case the damage was particularly severe near the tube support plates which are thought to have acted as heat conductors.

REMEDIAL MEASURES: As the operational conditions could not be altered the condenser was retubed with titanium.

COMMENT: Copper-nickel alloys of the 70:30 type are particularly prone to hot spot corrosion and therefore should not be used in circumstances where such conditions arise.

Figure 2: Section through a pit showing copper deposit.

Figure 1: Sectioned tube showing localised pitting of cooling water surface.

CASE HE 22

TUBES; ALUMINIUM BRASS; PITTING

INSTALLATION: Steam condenser in a power station.

DAMAGE: Pitting and perforation of a few tubes.

MATERIAL: The tubes were made to DIN 1795 (aluminium brass).

CONDITIONS: This unit was cooled by seawater and a few tubes perforated after 10 months in service.

It seems probably that such materials decomposed to produce sulphides which then reacted with the copper alloy to produce very cathodic films (> 200 mV with respect to base metal) on the tube surface. Rapid pitting then occurred at defects in these copper sulphide films.

The cooling water contained less than the detectable limit of sulphide (i.e. < 10µg/kg). Ferrous sulphate had been added to the cooling water and sponge rubber balls were circulated around the unit to keep tubes clean.

CAUSE OF DAMAGE: Most of the tubes were in good condition and were coated with a homogeneous light brown deposit. The pitted and perforated tubes did not have this appearance in that the deposits were multicoloured (green and brown) Figs. 1 and 2. In addition, these tubes contained plugs of organic materials such as dead fish, sea weed, mussels and algae.

REMEDIAL MEASURES: As only a few tubes were involved they were removed from service by plugging.

COMMENT: Heat exchanger tubes should not be allowed to become blocked and this can usually be prevented by ensuring adequate filtration. Biofouling of the cooling system may be prevented by the use of either electrochlorination or other techniques.

Figure 1: Sections through tubes showing patchy surface films .

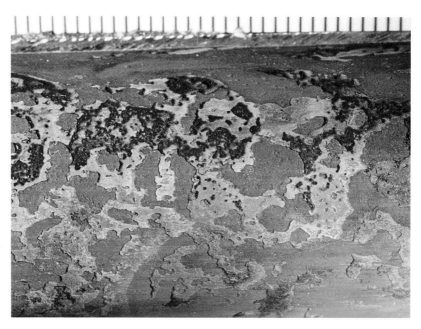

Figure 2: As above - greater detail.

CASE HE 23

TUBES; STAINLESS STEEL; PITTING

INSTALLATION: Seawater cooled heat exchanger in a chemical plant.

DAMAGE: Pitting and perforation of a number of tubes after one year of operation (see Fig.).

MATERIAL: The tubes were made to AISI 316.

CONDITIONS: The heat exchanger was cooled by sea water at ambient temperature. The temperature on the shell side was 40°C.

CAUSE OF DAMAGE: The presence of deposits (macrofouling) within the tubes caused the formation of crevices and this resulted in deep localized attack beneath the deposits.

REMEDIAL MEASURES: It may be possible to avoid this type of problem by an increased flow velocity in combination with mechanical cleaning of the tubes. Choice of a more corrosion resistant material is a more reliable approach. Highly alloyed stainless steels, copper alloys or titanium may be used.

COMMENT: Pitting and crevice corrosion are typical for conventional stainless steels like AISI 304 and AISI 316 in seawater. Cathodic protection from e.g. large carbon steel surfaces, may give satisfactory results, especially if the stainless steel surface is free from deposits.

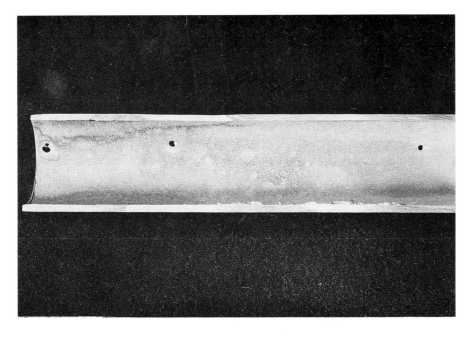

Pitting in AISI 316.

CASE HE 24

TUBES; ALUMINIUM BRASS; DEZINCIFICATION

INSTALLATION: Seawater cooled steam condenser in a power station.

DAMAGE: Numerous tube leaks occurred. When a tube was removed from the condenser, after approximately 45,000 hours in service, it disintegrated as shown in the Fig.

MATERIAL: The tubes were made from aluminium brass.

CONDITIONS: The condenser was cooled by seawater flowing at a nominal velocity of 1.8 m/s.

CAUSE OF DAMAGE: Severe dezincification of several tubes occured resulting in leaks. Chemical analysis of the failed tubes revealed that the arsenic concentration was rather low (approx. 0.02%). In addition, the magnesium content of the failed tubes was about 0.018% which was higher than the unfailed tubes (0.009%).

It is considered that this combination of relatively low arsenic and high magnesium content in the failed tubes is the reason for the observed dezincification.

REMEDIAL MEASURES: The damaged tubes were replaced with aluminium brass with more appropiate levels of arsenic and magnesium.

COMMENT: This type of failure has become less common since the causes of dezincification have been identified. The addition of arsenic is known to inhibit dezincification of alpha brasses and usually about 0.04% is present. However, it has been found that the presence of magnesium in the brass can prevent arsenic having its inhibiting effect.

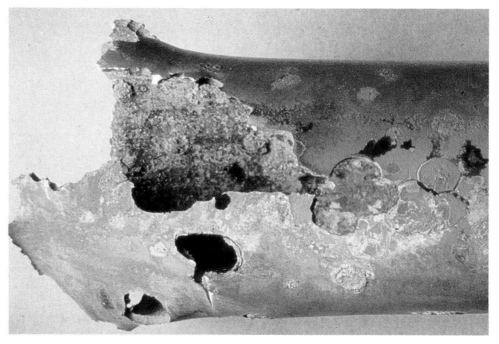

Aluminium brass tube damaged by dezincification.

CASE HE 25

TUBES; ADMIRALTY BRASS; STRESS CORROSION CRACKING

INSTALLATION: Steam condenser in a power plant installation cooled by brackish water.

DAMAGE: Cracking and leakage of condenser tubes.

MATERIAL: The tubes were made to DIN 1785 (admiralty brass).

CONDITIONS: The cooling water was brackish and polluted.

CAUSE OF DAMAGE: The damage, which is shown in Fig. 1 (below), and Figs.2 and 3 (shown on p. 114), is typical of that caused by stress corrosion cracking. The cracks have propagated from the cooling water side. As can be seen from the analysis given below the water was polluted:

O_2 = 3mg/kg
Chemical oxygen demand (COD)=10 mg/kg
NH_3 = 1 mg/kg

Stress corrosion cracking would not be expected to occur with this level of ammonia. It is considered that cracking occurred during stand-by periods when the tubes were filled with stagnant air and ammonia formed from decomposing organic deposits.

REMEDIAL MEASURES: The plant operating instructions were revised to ensure the avoidance of stagnant periods.

COMMENTS: Stress corrosion cracking of condenser tubes generally arises in the vicinity of areas of high stress such as roller expanded joints or local mechanical damage.

In the present case the source of high stresses is unknown.

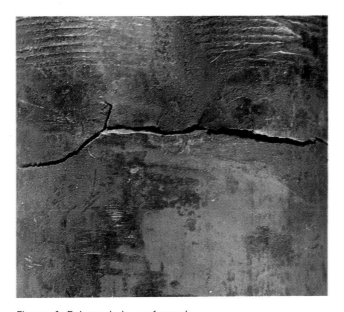

Figure 1: External view of crack.

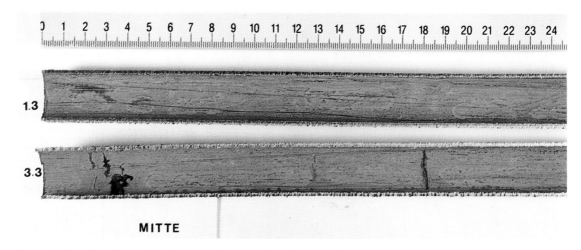

MITTE

Figure 2: Section through tube showing cracking from cooling water side.

Figure 3: Greater detail of above.

CASE HE 26

TUBES; ADMIRALTY BRASS; STRESS CORROSION CRACKING

INSTALLATION: 500 MW river water cooled steam condenser in a power station.

DAMAGE: Through wall stress corrosion cracking adjacent to the roller expanded tube to tube plate joints (Fig. 1).

MATERIAL: The tubes were made to BS 2871-CZ111 (Admiralty brass).

CONDITIONS: The condenser was cooled by river water flowing at a nominal velocity of 2 m/s.

CAUSE OF DAMAGE: This condenser was subject to an unexpected long downtime period. The condenser was not adequately drained and so stagnant water conditions arose due to decaying organic matter. This, in combination with the high residual tensile stresses in the material caused by roller expanding, resulted in stress corrosion cracking.

In this alloy the cracks follow a transgranular path (Fig.2).

REMEDIAL MEASURES: As a short term measure tube end inserts were fitted to shield the locally highly stressed area from the environment.

Subsequently the condenser was retubed with 90:10 copper nickel alloy which is resistant to cracking under these conditions.

COMMENTS: The basic problem is the development of an aggressive environment and thus stagnant water conditions should be avoided (or minimised). When downtimes occur the condenser should be thoroughly flushed in clean water and adequately drained.

Figure 2: Section through tube showing fine crack.

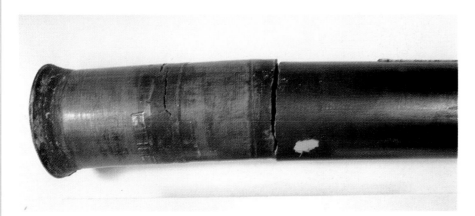

Figure 1: View of tube end showing cracking.

CASE HE 27

PLATE HEAT EXCHANGER; ALUMINIUM BRASS; STRESS CORROSION CRACKING

INSTALLATION: Ships lubricating oil plate heat exchanger cooled by seawater.

DAMAGE: Leakage of plates due to cracking and perforation.

MATERIAL: The plates were made to BS 2871-CZ110 (aluminium brass). They had been painted on one side with an epoxy type of paint cured with amines.

CONDITIONS: Seawater was present in the heat exchanger either stagnant or flowing at 90 $m^3 h^{-1}$ for about equal times during 4 months of service. During flow the seawater temperatures were typically inlet 22°C, outlet 28°C, with corresponding lubricating oil temperatures of 48°C and 40°C.

CAUSE OF DAMAGE: Stress corrosion cracking had originated at the convex region of the dimples (see Fig.).

In these areas the hardness of the alloy was 98 HV5 compared with a recommended maximum of 75 HV5 which is usual for this material in the fully annealed condition.

This indicates that the material has been stressed and when an ammonia test (BS 2871 part 3, 1972) was applied to portions of the plates cracking occurred indicating that they were susceptible to stress corrosion cracking. In service the corroding media could have originated from either the stagnant seawater or the amine in the paint (it was claimed that the paint would be beneficial during precommissioning).

REMEDIAL MEASURES: Plates replaced with fully annealed, uncoated material.

COMMENTS: In addition to avoiding stagnant conditions (see CASE HE 26) painted plates should not be used since at the very least they will contaminate the cooler later in service and may cause pitting on the water side.

Surface cracking on plate adjacent to perforation (etched in alcoholic ferric chloride).   x250

CASE HE 28

TUBES; ADMIRALTY BRASS; STRESS CORROSION CRACKING

INSTALLATION: Condenser in a refrigeration compressor cooled by brackish water.

DAMAGE: Cracking and leakage of condenser tubes.

MATERIAL: The tubes, which were finned, were made to BS 2871 part 3 CZ111 (Admiralty brass).

CONDITIONS: The condenser contained a Freon refrigerant and was cooled by brackish river water which could stagnate in the tubes whilst the unit was not in use. The failed tubes had been operative for 2500 hours during 6.5 years life of the unit.

CAUSE OF DAMAGE: A loss of refrigerant in the air-conditioning plant was traced to leaking tubes in a condenser. In order to determine whether complete retubing of this condenser was required several tubes were examined.

The tubes were found to contain transgranular cracks (see Fig.) propagating from the water side. This damage was attributed to stress corrosion cracking which occurred during shut down periods. Whilst the exact nature of the corrodant is not known it is believed to be associated with the decomposition of organic material present in the water.

REMEDIAL MEASURES: The unit was retubed and it was suggested that in the future, prior to extended shut downs, all condenser tubes should be brush cleaned internally and then filled with clean water.

Section through tube showing cracking.

CASE HE 29

TUBES; ALUMINIUM BRASS; FATIGUE

INSTALLATION: 500 MW seawater cooled steam condenser in a power station.

DAMAGE: Circumferential cracks appeared in a number of tubes near to the edge of the bundle. The failures occurred shortly after modification to the steam baffles had been made.

MATERIALS: The tubes were made to BS 378-CZ110 (aluminium brass).

CONDITIONS: The condenser was cooled by seawater flowing at a nominal velocity of 2 m/s.

CAUSE OF DAMAGE: Cracks, of the type shown in Fig. 1, had occurred midway between the tube support plates. The modifications carried out in the steam space had altered the steam flow characteristics in the region mentioned. This had caused the tubes to fail by a fatigue mechanism.

REMEDIAL MEASURES: Mid span anti-vibration devices, such as shown in Fig. 2, were fitted to the tubes adjacent to the affected area.

COMMENTS: Fatigue failures of heat exchanger tubes, which are not very common, are usually associated with design problems. When they occur it is usually very soon after commissioning.

Figure 1: Tube showing crack.

Figure 2: Anti vibration strips fitted between tube layers in a condenser.

CASE HE 30

TUBES; ALUMINIUM BRASS; FATIGUE

INSTALLATION: 500 MW seawater cooled steam condenser in a power station.

DAMAGE: Circumferential crack in a tube.

MATERIAL: The tube was made to BS 378-CZ110 (aluminium brass).

CONDITIONS: The condenser was cooled by seawater flowing at a nominal velocity of 2 m/s.

CAUSE OF DAMAGE: The circumferential crack had originated at a defect in the tube surface (see Fig.) and had propagated by a fatigue mechanism. The defect was found to be the result of an accidental arc struck by a welder during some internal condenser modifications to the steam baffle steel work.

REMEDIAL MEASURES: The leaking tube was plugged.

COMMENTS: Tube failures due to incidents such as this are extremely uncommon.

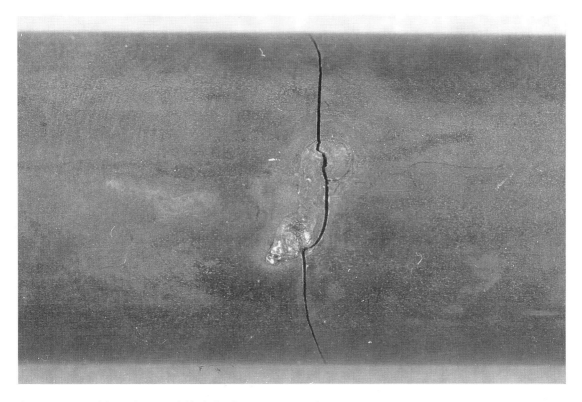

Crack initiated from "arc weld" defect.

CASE HE 31

TUBES; ALUMINIUM BRASS; FRETTING/FATIGUE

INSTALLATION: 150 MW seawater cooled
condenser in a power plant.

DAMAGE: Where the tubes had been fitted
into the support plates they had been damaged
by fretting which had resulted in wall thinning
followed by longitudinal cracking.

MATERIAL: The unit was tubed with material
conforming to DIN 1795 (aluminium brass).

CONDITIONS: The tubes were cooled by
seawater flowing at a nominal velocity of
2 m/s.

CAUSE OF DAMAGE: The failed tubes were in
the first row adjacent to the steam inlet duct.
Failure occurred only 3 months after retubing.
It was found that by mistake the first row had
been tubed with tubes of the same size as the
rest of the condenser. However, the old tubes
in this location were of larger diameter and
heavier wall thickness.

Vortex shedding, due to high velocity steam,
had caused the tubes to vibrate thus causing
fretting of the tube in the support plate (Fig. 1).
A longitudinal fatigue crack had developed as
the result of fretting and subsequent
alternating stressing of the tube in this area
(Figs. 2 and 3).

REMEDIAL MEASURES: The tubes affected were
plugged from the water side and in order to
limit vibrations, stakes were fitted to the steam
side.

COMMENTS: Tube failures due to such
incidents are very uncommon.

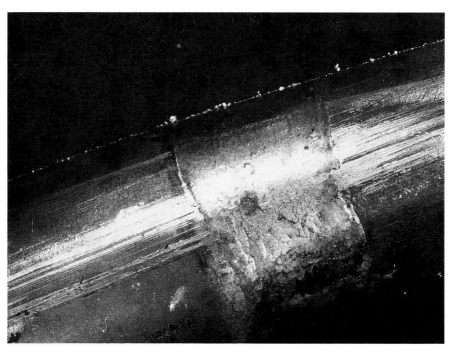

Figure 1: An aluminium brass tube damaged by fretting, within the support
plate x2.

Figure 2: Longitudinal fatigue crack that had developed from the fretting damage x2.

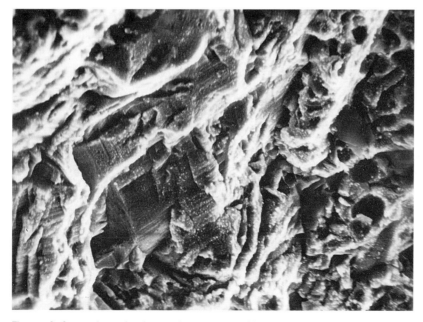

Figure 3: Scanning electron micrograph of the fracture surface showing striations characteristic of fatigue.

CASE HE 32

TUBES; ALUMINIUM BRASS; CAVITATION/WEAR

INSTALLATION: Main condenser of a 350,000 dwt oil tanker.

DAMAGE: Severe wear on the steamside and pitting damage on the waterside opposite to the wear.

MATERIAL: The tubes were made to BS 378-CZ110 (aluminium brass).

CONDITIONS: The unit is seawater cooled with the water flowing at a design velocity of 1.8 m/s.

CAUSE OF DAMAGE: Shortly after commissioning problems were experienced with the condenser due to excessive tube vibration which caused some tubes to rub against others.

The tubes were severely thinned due to a combination of wear on the steamside and cavitation/erosion on their watersides (Figs. 1 and 2). Both types of damage had been caused at the same time by excessive tube vibration.

COMMENTS: Fatigue/wear failures of heat exchanger tubes, which are not very common, are usually associated with design problems. When they occur it is usually very soon after commissioning.

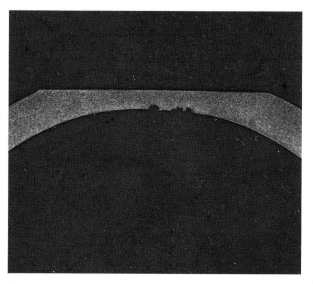

Figure 1: Section through tube wall showing external wear (flattened surface).

REMEDIAL MEASURES: Further damage was prevented by (a) strapping and wedging the tubes together, which prevented further cavitation and wear damage, and (b) treating the seawater with ferrous sulphate to increase the amount of iron in the system to help in the formation of protective films and to reduce significantly the probability of cavitation/erosion damage.

Figure 2: Section through tube showing cavitation/erosion on cooling water side.

CASE HE 33

TUBES; ALUMINIUM BRASS; STEAMSIDE CORROSION

INSTALLATION: Main condenser on a 250,000 dwt oil tanker.

DAMAGE: External corrosion and perforation of tubes (see Fig.).

MATERIAL: The tubes were made to BS 378-CZ110 (aluminium brass).

CONDITIONS: The tubes were cooled by seawater flowing at a nominal velocity of 1.8 m/s.

CAUSE OF DAMAGE: The damage was most severe on tubes either in or adjacent to the air extraction section and near to support plates. The damage was associated with the use of an excess of hydrazine, an oxygen scavenger, which decomposes to produce ammonia. This problem is often associated with slow steaming as during such operation hydrazine levels are difficult to control. The presence of carbon dioxide and oxygen has an accelerating effect upon the corrosion process.

REMEDIAL MEASURES: The tubes in the air extraction section of the condenser were replaced with tubes made from a copper-nickel alloy.

COMMENTS: In some cases the incidence of failures, due to this cause, have been reduced by applying closer control to the hydrazine levels within the system.

Tube showing severe external corrosion.

CASE HE 34

TUBES; ARSENICAL BRASS; STEAMSIDE CORROSION

INSTALLATION: River water cooled steam condenser in a power station.

DAMAGE: External general corrosion and perforation of tubes (see Fig. 34) near the tube plate.

MATERIAL: The tubes were made to BS 378 (70:30 arsenical brass)

CONDITIONS: The tubes were cooled by river water flowing at a nominal velocity of 2 m/s.

CAUSE OF DAMAGE: This type of corrosion is caused by the presence of ammonia on the steam side. Hydrazine, which is added to the boiler feed water to remove oxygen, decomposes to form ammonia which then condenses onto the tubes. The damage is intensified in the presence of ammonia, carbon dioxide and oxygen and is usually most severe adjacent to the tube support plates.

REMEDIAL MEASURES: Local retubing was carried out using copper-nickel alloy which is resistant to such attack.

COMMENTS: The close control of hydrazine in feed water is important particularly in systems using brass heat exchanger tubing.

Tubes showing attack and perforation.

CASE HE 35

TUBES; ALUMINIUM BRASS; STEAMSIDE EROSION

INSTALLATION: Main condenser on a 250,000 dwt oil tanker.

DAMAGE: External pitting erosion (see Fig.).

MATERIAL: The tubes were made to BS 378-CZ110 (aluminium brass).

CONDITIONS: The tubes were cooled by seawater flowing at a nominal velocity of 1.8 m/s.

CAUSE OF DAMAGE: The tubes affected by pitting were those adjacent to the steam inlet ducts. The damage was due to direct impingement of water droplets present in the steam.

REMEDIAL MEASURES: In this particular instance this damage did not present a major operational problem in that it did not result in excessive thinning or perforation.

COMMENTS: If the damage due to this cause does become severe the problem may be overcome by either using more erosion-resistant alloys or by fitting non-metallic erosion resistant shields locally. The possibility of plugging the affected tubes and using these as shields should be considered.

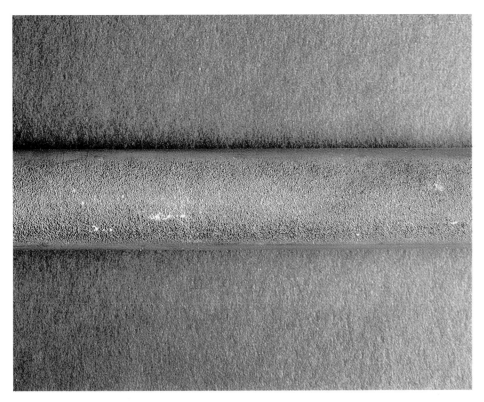

Tube showing pitting on external surface.

CASE HE 36

TUBES; TITANIUM; STEAMSIDE EROSION

INSTALLATION: Seawater cooled steam condenser in a power station.

DAMAGE: External pitting erosion (Figs. 1, 2 and 3). The tubes affected were those adjacent to the steam inlet ducts. In this instance about 50% of the tube wall thickness had been removed (Fig. 3).

MATERIAL: The tubes were made from titanium.

CONDITIONS: The tubes were cooled by seawater flowing at a nominal velocity of 2 m/s.

CAUSE OF DAMAGE: This damage was due to the impingement of water droplets present in the steam onto the titanium tube surface.

REMEDIAL MEASURES: As in this instance only a few tubes were affected they were plugged off.

COMMENTS: As in the previous case the damage may be prevented by fitting non-metallic erosion resistant shields locally.

Figure 1: Tube showing external erosion.

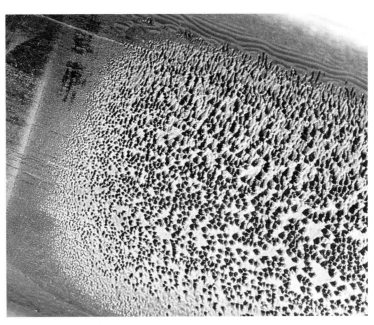

Figure 2: As above - greater detail.

Figure 3: Section through tube wall.

CASE HE 37

TUBES; COPPER-NICKEL; STEAMSIDE EROSION

INSTALLATION: Estuarine water cooled steam condenser in a 300 MW power station.

DAMAGE: External pitting and erosion which had led to perforation of tubes (see Fig.).

MATERIAL: The tubes were made to BS 2871 part 3, CN 108 (70:30 type copper-nickel).

CONDITIONS: The tubes were cooled by estuarine water flowing at a nominal velocity of 2 m/s.

CAUSE OF DAMAGE: The damage was caused by the direct impingement of water droplets onto the external surface of the condenser tubes and resulted in their erosion. The presence of water droplets in the steam is usually associated with either the mode of operation of the unit or its design. The damage is always most severe in the areas of the condenser most prone to water droplet formation.

REMEDIAL MEASURES: In this case plugging of the leading edge tubes was carried out.

COMMENTS: The fitting of erosion resistant non-metallic shields locally was an alternative solution.

External view of tube bundle showing area of damage.

CASE HE 38

PLATE HEAT EXCHANGER; TITANIUM; FRETTING CORROSION

INSTALLATION: Fresh water plate heat exchanger on a 350,000 dwt oil tanker.

DAMAGE: Perforation of one of the four blind corner discs that was positioned next to the steel end plate remote from the water inlet and outlet occured after about 1.5 years in service. The perforated area was a circular hole about 90 mm in diameter and the area around the hole was covered with rust (Figs. 1 and 2).

MATERIAL: Titanium.

CONDITIONS: The plate heat exchanger was cooled by seawater.

CAUSE OF DAMAGE: Operational pressures of about 3 bar had caused slight distortion to occur to the blind corner discs. This had allowed the titanium to make contact with the steel backing plate. The vibration on the vessel was then sufficient to initiate fretting corrosion of the titanium in a roughly circular manner. Once fretting had occurred thinning soon followed and when a hole was worn through the seawater penetrated to the steel backing plate causing it to rust.

It is more difficult to postulate the next step since in a galvanic couple between steel and titanium an increase in the corrosion of the steel would be expected with the titanium being protected. However, as fretting debris and rust had embedded into the naturally protective titanium oxide film it is possible for the titanium to corrode, since the oxide film would not be able to reform completely at damaged regions. This could allow exposed metal to occur.

Whilst the latter comment is somewhat speculative there is little doubt that distortion followed by fretting caused the failure.

REMEDIAL MEASURES: Replacement plates were fitted to the affected unit.

COMMENTS: This type of failure is extremely rare in titanium units.

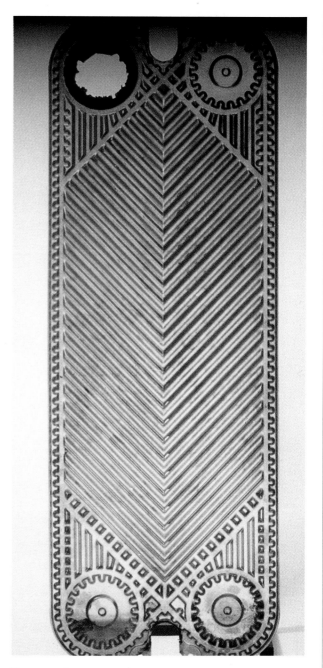

Figure 1: General view of damaged plate.

Figure 2: Close up of damaged plate showing perforated hole.